Lecture Notes in Computer Science 12293

More information about this series at http://www.springer.com/series/7411

Oussama Habachi · Vahid Meghdadi ·
Essaid Sabir · Jean-Pierre Cances (Eds.)

Ubiquitous Networking

5th International Symposium, UNet 2019
Limoges, France, November 20–22, 2019
Revised Selected Papers

Editors
Oussama Habachi (iD)
University of Limoges
Limoges, France

Vahid Meghdadi (iD)
University of Limoges
Limoges, France

Essaid Sabir (iD)
Hassan II University
Casablanca, Morocco

Jean-Pierre Cances (iD)
University of Limoges
Limoges, France

ISSN 0302-9743   ISSN 1611-3349 (electronic)
Lecture Notes in Computer Science
ISBN 978-3-030-58007-0   ISBN 978-3-030-58008-7 (eBook)
https://doi.org/10.1007/978-3-030-58008-7

LNCS Sublibrary: SL5 – Computer Communication Networks and Telecommunications

This Springer imprint is published by the registered company Springer Nature Switzerland AG
The registered company address is: Gewerbestrasse 11, 6330 Cham, Switzerland

Preface

UNet is an international scientific event that highlights new trends and findings in hot topics related to ubiquitous computing/networking. This 5th edition was held during November 20–22, 2019, in the fascinating city of Limoges, France.

Ubiquitous networks sustain development of numerous paradigms/technologies such as distributed ambient intelligence, Tactile Internet, context awareness, cloud computing, wearable devices, and future mobile networking (e.g., B4G, 5G, 6G). Various domains are then impacted by such a system, one cite security and monitoring, energy efficiency and environment protection, e-health, precision agriculture, intelligent transportation, home-care (e.g., for elderly and disabled people), etc. Communication in such a system has to cope with many constraints (e.g., limited capacity resources, energy depletion, strong fluctuations of traffic, real-time constraint, dynamic network topology, radio link breakage, interferences, etc.) and has to meet the new application requirements. Ubiquitous systems bring many promising paradigms aiming to deliver significantly higher capacity to meet the huge growth of mobile data traffic and to accommodate efficiently dense and ultra-dense systems. A crucial challenge is that ubiquitous networks should be engineered to better support existing and emerging applications including broadband multimedia, machine-to-machine applications, Internet of Things, sensor networks, and RFID technologies. Many of these systems require stringent quality of service including better latency, reliability, higher spectral, and energy efficiency, but also some quality-of-experience and quality-of-context constraints.

The main purpose of the UNet conference is to serve as a forum that brings together researchers and practitioners from academia and industry to discuss recent developments in pervasive and ubiquitous networks. The conference provides a forum to exchange ideas, discuss solutions, debate identified challenges, and share experiences among researchers and professionals. UNet also aims to promote the adoption of new methodologies and to provide the participants with advanced and innovative tools able to catch the fundamental dynamics of the underlying complex interactions (e.g., artificial intelligence, game theory, mechanism design theory, machine learning theory, SDR platforms, etc.).

Welcome Message from the UNet 2019 Chairs

It is our pleasure to welcome you to the proceedings of the 2019 edition of the International Symposium on Ubiquitous Networking (UNet 2019). The conference was held in the city of Limoges, France, during November 20–22, following up on the success of past editions. France has a prominent and active community of networking researchers and the choice of Limoges for UNet 2019 allowed its attendees, coming from all parts of the globe, to interact in a fascinating environment.

The growth of pervasive and ubiquitous networking in the past few years has been unprecedented. Today, a significant portion of the world's population is connected to the Internet most of the time through smart phones, and the Internet of Things promises to broaden the impact of the Internet to encompass devices ranging from electric appliances and medical devices to unmanned vehicles. The goal of UNet is to be a premier forum for discussing technical challenges and solutions related to such a widespread adoption of networking technologies, including broadband multimedia, 5G, Internet of Things, Tactile Internet, artificial intelligence for networking, security and privacy, data engineering, sensor networks, and RFID technologies. Toward this aim, we had five main technical tracks of papers covering all aspects of ubiquitous networks.

The UNet 2019 program featured four invited talks presented by distinguished keynote speakers: Prof. Jean-Claude Belfiore from Huawei Technologies (France), Prof. Sofie Pollin from Catholic University of Leuven (Belgium), Prof. Latif Ladid from University of Luxembourg (Luxembourg), and Prof. Catherine Douillard from IMT Atlantique (France). With a rich program that reflects the most recent advances in ubiquitous computing and ambient intelligence, involving a broad range of theoretical tools (e.g., game theory, mechanism design theory, learning theory, machine learning, etc.) and practical methodologies (e.g., SDR/SDN platforms, embedded systems, privacy and security by design, etc.) to study modern technologies (5G, Internet of Things, Tactile Internet, industry 4.0, etc.). We were very pleased to welcome our attendees to this new edition of the UNet conference series.

We are very thankful to the XLIM, ENSIL-ENSCI School of Engineering, and NEST Research Group for co-organizing this exciting event. We are grateful to our technical sponsor Springer Science+Business Media, without whom UNet 2019 would not have been possible. We are also very thankful to all our sponsors and patrons (ENSIL-ENSCI, University of Limoges, ENSEM, Hassan II University of Casablanca, and Maghreb Solutions).

Enjoy the proceedings!

November 2019

Oussama Habachi
Stefano Secci
Jean-Pierre Cances

Welcome Message from the UNet 2019 TPC Chairs

It is with great pleasure that we welcome you to the proceedings of the 2019 International Symposium on Ubiquitous Networking (UNet 2019), which was held in Limoges, France. The conference featured an interesting technical program of five technical tracks reporting on recent advances in ubiquitous communication technologies and networking, Tactile Internet and Ubiquitous Internet of Things, mobile edge networking and fog-cloud computing, AI and machine learning for ubiquitous communications, and data engineering, cyber security, and pervasive services. UNet 2019 also featured four keynote speeches by world-class experts, and one invited paper session.

We received 41 paper submissions from 14 countries and 4 continents. From these, 17 regular papers and 1 short paper were accepted after a careful review process to be included in the UNet 2019 proceedings. We also included two invited papers from acknowledged researchers. The regular-paper acceptance rate was 41% whereas the overall acceptance rate in UNet 2019 was 43%.

The preparation of this excellent program would not have been possible without the dedication and hard work of the different chairs, the keynote speakers, and all the Technical Program Committee members and reviewers. We take this opportunity to acknowledge their valuable work, and sincerely thank them for their help in ensuring that UNet 2019 will be remembered as a high-quality event.

We hope that you will enjoy this edition's proceedings.

November 2019

Bo Ji
Essaid Sabir
Vahid Meghdadi

Organization

General Chairs

Oussama Habachi	University of Limoges, XLIM-SRI, France
Stefano Secci	CNAM Paris, France

Local Chair

Jean-Pierre Cances	University of Limoges, Head of XLIM-SRI, France

TPC Chairs

Bo Ji	Temple University, USA
Essaid Sabir	ENSEM, Hassan II University of Casablanca, Morocco
Vahid Meghdadi	University of Limoges, XLIM-SRI, France

Track Chairs

Track #1 (Ubiquitous Communication Technologies and Networking) Chairs

Anne Julien-Vergonjanne	University of Limoges, XLIM-SRI, France
Majed Haddad	University of Avignon, France

Track #2 (Tactile Internet and Ubiquitous Internet of Things) Chair

Mehdi Bennis	University of Oulu, Finland

Track #3 (Mobile Edge Networking and Fog-Cloud Computing) Chair

Halima Elbiaze	Université du Québec à Montréal (UQAM), Canada

Track #4 (AI and Machine Learning for Ubiquitous Communications) Chairs

Tembine Hamidou	University of New York, USA/UAE
Bin Li	University of Rhode Island, USA

Track #5 (Data Engineering, Cyber Security and Pervasive Services) Chairs

Quanyan Zhu	New York University Tandon School of Engineering, USA
Loubna Echabbi	INPT, Morocco

Tutorials, Workshops and Special Sessions Chairs

Francesco De Pellegrini	University of Avignon, France
Piotr Wiecek	University of Wrocław, Poland

Publication Chairs

Abbas Bradei	University of Poitiers, France
Mohamed Sadik	ENSEM, Hassan II University of Casablanca, Morocco

Local Arrangement and Registration Chairs

Emmanuel Conchon	University of Limoges, France
Karim Tamine	University of Limoges, France

Publicity Chairs

Abdellatif Kobbane	ENSIAS, Morocco
Antonio Jara	University of Murcia, Spain
Damien Sauveron	University of Limoges, XLIM-SRI, France
Hajar El Hammouti	KAUST, Saudi Arabia
Somayyeh Chamaani	KNTU, Iran
Taher Ezzeddine	ENIT, Tunisia

Technical Program Committee

Mojtaba Aajami	Islamic Azad University, Zanjan Branch, Iran
Walid Abdallah	CN&S Research Lab, Tunisia
Abdelkrim Abdelli	USTHB University, Algeria
Bahareh Akhbari	K. N. Toosi University of Technology, Iran
Noura Aknin	Abdelmalek Essaâdi University, Morocco
Adel Al-Hezmi	T-Systems International GmbH, Germany
Eyhab Al-Masri	University of Washington, USA
Vangelis Angelakis	Linköping University, Sweden
Imran Shafique Ansari	University of Glasgow, UK
Marwane Ayaida	University of Reims Champagne-Ardenne, France
Elarbi Badidi	UAE University, UAE
Abdelmajid Badri	FSTM UH2C, Morocco
Luis Barbosa	University of Castilla La Mancha, Spain
Stylianos Basagiannis	United Technologies Research Centre, Ireland
Hicham Behja	ENSEM, Morocco
Giampaolo Bella	University of Catania, Italy
Paolo Bellavista	University of Bologna, Italy
Asma Ben Letaifa	SupCom, Tunisia
Yann Ben Maissa	INPT, Morocco
Salah Benabdallah	University of Tunis, Tunisia

Imade Benelallam	INSAE, Morocco
Mustapha Benjillali	INPT, Morocco
Yahya Benkaouz	FSR, Mohammed V University, Morocco
Fatma Benkhelifa	Imperial College London, UK
Hassan Bennani	ENSIAS, Mohammed V University, Morocco
Ana Bernardos	Universidad Politecnica de Madrid, Spain
Victoria Betran	Technical University of Madrid (UPM), Spain
Md Zakirul Alam Bhuiyan	Fordham University USA
Eleonora Borgia	IIT-CNR, Italy
Leila Boulahia	University of Technology of Troyes, France
Jaouad Boumhidi	University of Sidi Mohamed Ben Abdellah, Morocco
Olivier Brun	Laboratoire d'Analyse et d'Architecture des Systemes, France
Lin Cai	Illinois Institute of Technology, USA
Bin Cao	Harbin Institute of Technology, China
Stefano Chessa	Università di Pisa, Italy
Satish Chikkagoudar	US Naval Research Laboratory, USA
Alessandro Chiumento	University of Twente, The Netherlands
Domenico Ciuonzo	University of Naples Federico II, Italy
Hamza Dahmouni	INPT, Morocco
Sabrina De Capitani di Vimercati	Università degli Studi di Milano, Italy
Yacine Djemaiel	University of Carthage, Tunisia
Ciprian Dobre	University Politehnica of Bucharest, Romania
Schahram Dustdar	Vienna University of Technology, Austria
Loubna Echabbi	INPT, Morocco
Faissal El Bouanani	ENSIAS, Mohammed V University, Morocco
Hajar El Hammouti	KAUST, Saudi Arabia
Mohamed El Kamili	Hassan II University of Casablanca, Morocco
Ahmed El Maliani drissi	Faculté des Sciences, Université Mohammed V, Morocco
Mourad El Yadari	FPE, UMI, Morocco
Rachid El-Azouzi	University of Avignon, France
Oussama Elissati	INPT, Morocco
Larbi Esmahi	Athabasca University, Canada
Moez Esseghir	University of Technology of Troyes, France
Sidi Ahmed Ezzahidi	Mohamed V University, Morocco
Xinxin Fan	IoTeX, USA
Habib Fathallah	Carthage University, Tunisia
Gianluigi Ferrari	University of Parma, Italy
Mohamed Fezari	Badji Mokhtar Annaba University, Algeria
Dieter Fiems	Ghent University, Belgium
Rosa Figueiredo	University of Avignon, France
Stefan Fischer	University of Lübeck, Germany
Giancarlo Fortino	University of Calabria, Italy
Alexandros Fragkiadakis	ICS-FORTH, Greece

Miguel Franklin de Castro	Federal University of Ceará, Brazil
Vasilis Friderikos	King's College London, UK
Aminata Garba	Carnegie Mellon University, USA
Yacine Ghamri-Doudane	University of La Rochelle, France
Alireza Ghasempour	ICT Faculty, USA
Hicham Ghennioui	LSSC, University of Sidi Mohammed Ben Abdellah, Morocco
Mounir Ghogho	International University of Rabat, Morocco
Gaetano Giunta	University of Roma Tre, Italy
Stefanos Gritzalis	University of Piraeus, Greece
Oussama Habachi	XLIM, France
Majed Haddad	University of Avignon, France
Yassine Hadjadj-Aoul	University of Rennes 1, France
Ridha Hamila	Qatar University, Qatar
Moulay Lahcen Hasnaoui	Université Moulay Ismail, Ecole Supérieure de Technologie, Morocco
Silkan Hassan	Université Chouaib Doukkali, Morocco
José Luis Hernandez Ramos	European Commission - Joint Research Centre (JRC), Belgium
Amal Hyadi	McGill University, Canada
Khalil Ibrahimi	University of IBN Tofail, Morocco
Muhammad Ali Imran	University of Glasgow, UK
Dimosthenis Ioannidis	Information Technologies Institute, Greece
Isam Ishaq	Al-Quds University, Palestine
Tawfik Ismail	NILES, Cairo University, Egypt
Frank Johnsen	Norwegian Defence Research Establishment (FFI), Norway
Carlos Kamienski	Universidade Federal do ABC, Brazil
Vasileios Karyotis	Ionian University, Greece
Donghyun Kim	Georgia State University, USA
Hyunbum Kim	University of North Carolina at Wilmington, USA
Jong-Hoon Kim	Kent State University, USA
Abdellatif Kobbane	ENSIAS, Mohammed V University, Morocco
Manel Kortas	University of Tunis El Manar, Tunisia
Mohamed Koubaa	Université Tunis El Manar, Ecole Nationale d'Ingénieurs de Tunis, Tunisia
Mohammed-Amine Koulali	University Mohammed I, Morocco
Rim Koulali	Faculté des Sciences Ain Chok, Morocco
Gyu Myoung Lee	Liverpool John Moores University, UK
Shancang Li	University of the West of England, UK
Marco Listanti	University of Rome La Sapienza, Italy
Jaime Lloret	Universitat Politecnica de Valencia, Spain
Diego Lopez	Telefonica I+D, Spain
Michael Losavio	University of Louisville, USA
Valeria Loscrí	Inria Lille-Nord, France
Malamati Louta	University of Western Macedonia, Greece

Dario Vieira	EFREI, France
Om Vyas	Indian Institute of Information Technology, Allahabd, India
Yunsheng Wang	Kettering University, USA
Wei Wei	Xi'an University of Technology, China
Konrad Wrona	NATO Communications and Information Agency, The Netherlands
Yang Xiao	The University of Alabama, USA
Chengwen Xing	Beijing Institute of Technology, China
Li Xu	Fujian Normal University, China
Liang Yang	Hunan University, China
Mariem Zayene	University of Tunis El Manar, Tunisia
Sherali Zeadally	University of Kentucky, USA
Emna Zedini	KAUST, USA
Jie Zeng	Tsinghua University, China
Ping Zhou	Apple, USA
Zbigniew Zielinski	Military University of Technology, Poland

UNet 2019 Keynote Speakers

From Learning to Reasoning: A Topos Perspective

Jean-Claude Belfiore

Abstract. With the explosion of data, the evolution of 5G networks towards 2020–2030 will be mostly based on machine learning techniques. They provide the first era of intelligent networks which we call learning networks. The second era of beyond 5G intelligent network for the period 2030–2040 are networks which are able to think. This requires new advanced mathematical tools which go beyond the perceptual framework of machine learning now. This can be done by linking the topology of perception and the logic of thinking, using category theory and its deeper notion of topos, invented by Alexandre Grothendieck at IHES in the 60s. As a bonus, this intriguing connection makes the notion of semantics appear naturally. Can we finally now start to build the foundations of semantic communication evoked by Shannon and Weaver in the early 50s?

Jean-Claude Belfiore graduated from Supelec, France, received a PhD from Télécom ParisTech, and the Habilitation (HdR) from Université Pierre et Marie Curie (UPMC). Until 2015, he was with Télécom ParisTech as a Full Professor in the Communications and Electronics Department. In 2015, he joined the Mathematical and Algorithmic Sciences Lab of Huawei.

Jean-Claude Belfiore has made pioneering contributions on modulation and coding for wireless systems (especially space-time coding) by using tools of number theory. He is also one of the co-inventors of the celebrated Golden Code of the Wi-Max standard.

Jean-Claude Belfiore is author or co-author of more than 200 technical papers and communications and has served as advisor for more than 30 PhD students.

He was Associate Editor of the *IEEE Transactions on Information Theory for Coding Theory* and has been the recipient of the 2007 Blondel Medal.

Since November 2015, Jean-Claude Belfiore has joined the Paris Research Center of Huawei Technologies where he leads a department. He has been involved in the 5G standardization process, essentially for the channel coding track (polar codes for 5G).

He now participates in the definition of 6G. Among other areas at Huawei, he is also actively related to the foundations of artificial intelligence.

Electrosense, Open and Big Spectrum Data

Sofie Pollin

Abstract. With the explosion of wireless devices, there is a growing number of applications that require a deep understanding of the actual spectrum usage. New technologies are needed that go beyond or can complement classical high-end spectrum analyzers. Electrosense is the first initiative that exploits the paradigms of low-cost programmable spectrum sensors, crowdsourcing to users, and big data architecture to gather and make available spectrum data and events to scientists, practitioners, and stakeholders. In this talk we will review the main design concepts of the Electrosense network, the main research findings, and how the scientific community can contribute to the network.

Sofie Pollin obtained her PhD degree at KU Leuven, Belgium, with honors in 2006. From 2006–2008 she continued her research on wireless communication, energy-efficient networks, cross-layer design, coexistence, and cognitive radio at UC Berkeley. In November 2008 she returned to imec to become a principal scientist in the green radio team. Since 2012, she is tenure track Assistant Professor at the Electrical Engineering Department at KU Leuven. Her research centers around networked systems that require networks that are ever more dense, heterogeneous, battery powered, and spectrum constrained. Prof. Pollin is a BAEF and Marie Curie fellow, and an IEEE Senior Member.

IPv6-Based Internet Empowering Super IoT, 5G and Blockchain while Cybersecurity is Looming

Latif Ladid

Abstract. The recent Mckinsey report on IoT projects 3 to 11 trillion dollars of IoT business by 2025. IoT is just the Internet sneaking everywhere. The current deployment of IoT is run over NAT converted logically to InterNAT of Things. The Hackers cannot wait to go after small fish to take down networks for money. 4G deployed over NAT except top notch ISPs like T-Mobile in the US using IPv6 and serving v4 customers with v4 as a service, or due to simply a lack of v4 address space such as in India with Reliance Jio deploying 4G with IPv6 and capturing 250 million 4G users, basically demonstrating a great case of a greenfield scenario leapfrogging developing countries into use of IPv6 without even knowing it. Blockchain is hailed to save the planet with its security. Again, the usual hype hits the road as blockchain is based on IPv4/NAT and the July 2018 hack of the keys got Bitcoin stumbling from 20 thousand dollars to 5 thousand dollars. However, people are not aware of these issues and keep speculating with Bitcoin, and some have even lost their keys, as is the case for one who lost 75 million dollars. This digital coin is not made for a layman. This talk will restore some sanity by looking at the historical developments of these technologies and learn from past mistakes and mind-boggling hypes.

Latif Ladid is a Senior Researcher at the Interdisciplinary Centre for Security, Reliability and Trust (SnT), Luxembourg. As a member of Secan-Lab, he works on multiple European Commission Next Generation Technologies IST Projects, including: 6INIT, www.6init.org – First Pioneer IPv6 Research Project; 6WINIT, Euro6IX, www.euro6ix.org; Eurov6, www.eurov6.org; NGNi, www.ngni.org; project initiator of SEINIT, www.seinit.org; and SecurIST, www.securitytaskforce.org.

Latif initiated the new EU project u-2010 to research Emergency & Disaster and Crisis Management, www.u-2010.eu; relaunched the Public Safety Communication Forum, www.publicsafetycommunication.eu; supported the new IPv6++ EU Research Project called EFIPSANS, www.efipsans.org, as well as the new Safety & Security Project using IPv6 called Secricom, www.secricom.eu; and co-initiated the new EU Coordination of the European Future Internet Forum for Member States called ceFIMS, www.ceFIMS.eu.

He holds the following positions: President, IPv6 FORUM, www.ip6forum.org; Chair, European IPv6 Task Force, www.ipv6.eu; Emeritus Trustee, Internet Society www.isoc.org; and Board Member IPv6 Ready & Enabled Logos Program and Board Member World Summit Award, www.wsis-award.org. Latif is also a Member of 3GPP PCG (www.3gpp.org), 3GPP2 PCG (www.3gpp2.org), Vice Chair, IEEE ComSoc EntNET (www.comsoc.org/~entnet/), Member of UN Strategy Council, Member of IEC Executive Committee, Board Member of AW2I, Board Member of Nii Quaynor Institute for Research in Africa, and Member of the Future Internet Forum EU Member States, representing Luxembourg: http://ec.europa.eu/information_society/activities/foi/lead/fif/index_en.htm.

Channel Coding for Tb/s Wireless Communications: Insights into Code Design, Decoding Algorithms and Implementation

Catherine Douillard

Abstract. While the wireless world is moving towards the 5G era, wireless Tb/s communications are expected to become a main technology trend within the next 10 years and beyond. On another note, for several decades, improvement in silicon technology has provided higher frequency, lower cost per gate, higher integration density, and lower power consumption. However, microelectronics has now reached a point where it can no longer keep pace with the increasing requirements of communication systems, alone. Therefore, the Tb/s data rate is a significant challenge for the design of transceivers and in particular for forward error correction, the most complex component in the baseband chain. Consequently, silicon implementations of advanced channel coding schemes require a cross-layer approach involving information theory, algorithm development, parallel hardware architectures, and semiconductor technology. This paper deals with the implementation challenges for advanced channel coding techniques, such as turbo codes, low-density parity-check (LDPC) codes or polar codes, when Tb/s throughput is targeted. As an example, we demonstrate how the specific design of codes and decoding algorithms, as well as the development of parallel hardware architectures make it possible to achieve a throughput higher than 100 Gb/s with current semiconductor technology.

Dr. Catherine Douillard received the engineering degree in telecommunications from the École nationale supérieure des télécommunications de Bretagne, France, in 1988, a PhD degree in electrical engineering from the University of Western Brittany, France, in 1992, and the accreditation to supervise research from the University of Southern Brittany, France, in 2004.

She is currently a full Professor in the Electronics Department of IMT Atlantique where she is in charge of the Algorithm-Silicon Interaction research team of the Lab-STICC laboratory. Her main research interests are error correcting codes, iterative decoding, iterative detection, coded modulations, and diversity techniques for multi-carrier, multi-antenna, and multiple access transmission systems.

Between 2007 and 2012, she participated in DVB (Digital Video Broadcasting) Technical Modules for the definition of DVB-T2, DVB-NGH, and DVB-RCS NG standards. She also served as the Technical Program Committee (co-)chair of ISTC 2010 and ISTC 2018 (International Symposium on Topics in Coding), as the general chair of ISTC 2016, and she will serve as the general co-chair of ISTC 2020.

In 2009, she received the SEE/IEEE Glavieux Award for her contribution to standards and related industrial impact.

Contents

Ubiquitous Internet of Things

Pervasive Services and Applications

Ubiquitous Communication Technologies and Networking

Comparison of Multi-channel Ranging Algorithms for Narrowband LPWA Network Localization

Florian Wolf[1,2](✉), Mohamed Sana[1](✉), Sébastien de Rivaz[1](✉),
François Dehmas[1](✉), and Jean-Pierre Cances[2](✉)

[1] CEA-Leti Minatec Campus, 17 rue des Martyrs, 38054 Grenoble Cedex 09, France
{florian.wolf,mohamed.sana,sebastien.derivaz,francois.dehmas}@cea.fr
[2] Université de Limoges, CNRS, XLIM, UMR 7252, 87000 Limoges, France
{florian.wolf,jean-pierre.cances}@xlim.fr

Abstract. Accurate radio signal based localization for Low Power Wide Area networks enables ubiquitous positioning for the Internet of Things. Narrowband communication and multipath propagation make precise localization challenging. Coherent multi-channel ranging increases bandwidth and provides improved temporal resolution through the aggregation of sequentially transmitted narrowband signals. This paper applies parametric estimators as well as a deep learning technique to multi-channel measurements obtained with 10 kHz signals. Ranging performances are compared via numerical simulations and real outdoor field trials, where parametric estimation and deep learning achieve 60 m and 45 m accuracy in 90% of the cases, respectively. Further work is required to study the impact of deep neural network training with a combination of synthetic and real data. Future research may also include the adaptation of multi-channel localization to differential network topologies.

Keywords: LPWA network localization · Range estimation ·
Frequency hopping · MUSIC · Deep learning · DNN

1 Introduction

Narrowband Low Power Wide Area (LPWA) networks based on radio technologies such as LoRa, Sigfox and NB-IoT provide wireless connectivity to low data rate devices in the framework of the Internet of Things (IoT) [1].

The localization of these devices enables new applications such as object tracking, allows adding context information to device-generated data or can be used for enhanced network management [2]. While integrating a Global Navigation Satellite System (GNSS) module ensures meter-level positioning, certain use cases exclude this option due to power, form factor or cost constraints. Extracting location information from LPWA radio signals carrying the transmitted data addresses these issues. However, obtaining precise position estimates from low data rate and narrowband signals, required for long-range communication over

© Springer Nature Switzerland AG 2020
O. Habachi et al. (Eds.): UNet 2019, LNCS 12293, pp. 3–17, 2020.
https://doi.org/10.1007/978-3-030-58008-7_1

several kilometers, remains challenging [3]. Low temporal resolution results in the difficulty to accurately estimate the Time-of-Arrival (ToA) of the direct path in multipath scenarios.

A brief literature review on LPWA network ranging and localization techniques is given in the following. In a mesh network, range estimates are obtained with a parametric signal path loss model and Received Signal Strength (RSS) measurements [4]. Due to blockage and small-scale fading, it is difficult to determine a valid model for RSS based ranging. To account for these shortcomings, RSS measurements combined with fingerprinting algorithms [5] achieve, once a robust database has been established, accuracy up to 500 m. Precision of ToA based ranging techniques is inverse proportional to signal bandwidth [6]. LPWA networks implementing Time-Difference-of-Arrival (TDoA) techniques obtain location estimates through hyperbolic trilateration with a 500 m precision [7].

Coherent multi-channel ranging [6] aggregates multiple sequentially transmitted narrowband signals on different frequencies to form a virtual increased bandwidth. This technique, similar to stepped frequency radar, is compatible with LPWA transceivers and improves temporal resolution [8]. Various implementations [9–12] have demonstrated a higher accuracy over single-channel time based ranging techniques. However, these studies have focused on short-range technologies (i.e. WiFi, Zigbee, RDIF) and mostly on indoor propagation scenarios.

Multi-channel range estimation is based on a sampled version of the channel transfer function. The delay estimation problem can be addressed through MUltiple SIgnal Classification (MUSIC) [13], allowing an improved resolution of multipath over Fourier transformation based techniques [14,15].

Recent advances in machine learning for wireless communications have opened up new horizons for the design of more flexible algorithms [16]. For range estimation problems, some of these algorithms are based upon Deep Neural Network (DNN) [17] or Convolutional Neural Network (CNN) [18] and are trained either using synthetically generated data or real-life measurements. The resulting solutions are shown to be competitive with respect to the state-of-art approaches while being flexible, as they comprise straightforward calculations [17].

The contributions of this paper, on coherent multi-channel ranging, are:

- Formulation of the multi-channel ranging signal model for the high-resolution algorithm MUltiple SIgnal Classification (MUSIC). Comparison of the direct channel impulse reconstruction by Inverse Discrete Fourier Transform (IDFT) and a MUSIC based range estimator in a two-path propagation scenario by numerical simulations.
- Design of a Deep Neural Network (DNN) based regression algorithm for range estimation, based on synthetically generated multi-channel transfer function data. Ranging performance comparison with IDFT and MUSIC by numerical simulations.
- Application and comparison of the parametric estimators (IDFT and MUSIC) as well as the deep learning algorithm (DNN) on real outdoor narrowband LPWA ranging field trials.

Range estimation through the IDFT technique i.e. as in [10] serves throughout this work as state of the art baseline approach in order to establish a fair comparison with MUSIC and DNN based ranging in the context of narrowband LPWA networks.

This paper is organized as follows. Section 2 introduces the multi-channel signal model and presents the range estimators. Comparison of the range estimation approaches is given by numerical simulation in Sect. 3. The multi-channel ranging transceiver testbed and the application of the range estimation techniques to real outdoor field trial measurements is discussed in Sect. 4. Section 5 concludes with perspectives.

2   Coherent Multi-channel Ranging

2.1   Signal Model

The basic transmission signal model including transmitter, radio channel and receiver is depicted in Fig. 1a. The energy normalized narrowband LPWA baseband signal s_0 is up-converted to transmit frequency f_0 resulting in the transmit signal

$$s_{\mathrm{TX}}(t) = \mathcal{R}\left\{ s_0\left(t^{[T]}(t)\right) e^{j\left(2\pi f_0 t^{[T]}(t) + \phi_{\mathrm{R}}^{[T]}\right)} \right\}, \tag{1}$$

where the transmitter local time $t^{[T]}(t)$ is function of global time t applying to both baseband signal and up-conversion operation and $\mathcal{R}(\cdot)$ is the real part. After passing through the radio channel, the received signal in global time t can be described by

$$s_{\mathrm{RX}}(t) = s_{\mathrm{TX}}(t) * h(t), \tag{2}$$

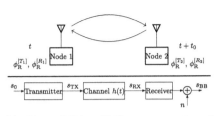

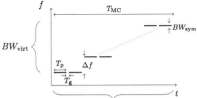

(a) Phase-of-Flight (PoF) measurement and transmission model.

(b) Ranging protocol aggregating C sequentially transmitted narrowband signals (BW_{sym}) to a virtual bandwidth $BW_{\mathrm{virt}} = (C-1)\Delta f$.

Fig. 1. Coherent multi-channel ranging.

with convolution operator $*$ and energy normalized channel impulse response, combining P multipath components

$$h(t) = \sum_{p=0}^{P-1} \alpha_p \delta(t - \tau_p) \quad \circ\!\!-\!\!\bullet \quad H(f) = \sum_{p=0}^{P-1} \alpha_p e^{-j2\pi f \tau_p}, \tag{3}$$

where $d = c_0 \tau_0$ is the direct path and speed of light $c_0 = 3 \cdot 10^8 \, \text{m/s}$. After down-conversion, the low pass filtered received baseband signal is given by

$$s_{\text{BB}}(t) = s_{\text{RX}}(t) \, e^{-j\left(2\pi f_0 t^{[R]}(t) + \phi_{\text{R}}^{[R]}\right)} + n^{[R]}(t), \tag{4}$$

where $t^{[R]}(t)$ is the receiver local time applying to the down-conversion operation only and Additive White Gaussian Noise (AWGN) n of variance $\sigma^2 = 1/(2E_{\text{s}}/N_0)$.

Combining (1), (2) and (4) results in the general received baseband signal

$$s_{\text{BB}}(t) = \left[s_0\left(t^{[T]}(t)\right) e^{j\left(2\pi f_0 t^{[T]}(t) + \phi_{\text{R}}^{[T]}\right)} * h(t) \right] e^{-j\left(2\pi f_0 t^{[R]}(t) + \phi_{\text{R}}^{[R]}\right)} + n^{[R]}(t). \tag{5}$$

In the following, a time offset t_0 between node 1 and node 2 and a zero Sampling/Carrier Frequency Offset (SFO/CFO) is assumed. Assigning node 1 as reference node, local to global time relations are expressed by

$$t^{[T_1]}(t) = t^{[R_1]}(t) = t, \tag{6a}$$
$$t^{[T_2]}(t) = t^{[R_2]}(t) = t + t_0. \tag{6b}$$

For a transmission from node 1 to node 2, the resulting received signal according to (5) and (6a), (6b) is given by

$$s_{\text{BB}}^{[T_1, R_2]}(t) = \left[s_0(t) \, e^{j\left(2\pi f_0 t + \phi_{\text{R}}^{[T_1]}\right)} * h(t) \right] e^{-j\left(2\pi f_0 (t+t_0) + \phi_{\text{R}}^{[R_2]}\right)} + n^{[R_2]}(t). \tag{7}$$

The frequency domain representation of (7), without noise n, can be written as

$$S_{\text{BB}}^{[T_1, R_2]}(f) = S_0(f) H(f + f_0) e^{-j2\pi f_0 t_0} e^{j\left(\phi_{\text{R}}^{[T_1]} - \phi_{\text{R}}^{[R_2]}\right)}. \tag{8}$$

Assuming normalization $S_0(f = 0) = 1$ and evaluating[1] (8) at $f = 0$, results in

$$H_c^{[T_1, R_2]} = \left. \left(S_{\text{BB}}^{[T_1, R_2]}(0) \right) \right|_{(f_0 = f_{\text{R}} + c\Delta f)} = H(f_0) e^{-j2\pi f_0 t_0} e^{j\left(\phi_{\text{R}}^{[T_1]} - \phi_{\text{R}}^{[R_2]}\right)}. \tag{9}$$

For time synchronized nodes i.e. $t_0 = 0$, a sampled estimation of the channel transfer function $H(f)$ can be obtained by performing narrowband transmissions from node 1 to node 2 according to Fig. 1b at frequencies $f_0 = f_{\text{R}} + c\Delta f$ with $c \in [0 \ldots C-1]$. This one-way transmission, generally called channel estimation, is not sufficient for range estimation between unsynchronized nodes ($t_0 \neq 0$)

---

[1] In practice, evaluation is performed by cross correlation with s_0 in time domain [8].

as the delay τ of the radio channel, i.e. $h(t) = \alpha\delta(t - \tau)$ and t_0 are linearly dependent. A two-way exchange with a transmission from node 2 to node 1 results in $s^{[T_2, R_1]}$. Equivalent to (9), the frequency domain representation is given by

$$H_c^{[T_2, R_1]} = H(f_0)e^{j2\pi f_0 t_0} e^{j\left(\phi_R^{[T_2]} - \phi_R^{[R_1]}\right)}. \tag{10}$$

Combining (9) and (10) eliminates the unknown time reference t_0

$$\tilde{H}_c = H_c^{[T_1, R_2]} H_c^{[T_2, R_1]} = H^2(f_R + c\Delta f)e^{j\left(\phi_R^{[T_1]} - \phi_R^{[R_2]} + \phi_R^{[T_2]} - \phi_R^{[R_1]}\right)}. \tag{11}$$

This sampled version of the channel transfer function serves as input to the range estimators discussed in Sect. 2.2.

2.2   Range Estimators

Three major issues need to be considered for range estimation based on (11):

- The sum-difference of local oscillator phases $\Delta\phi_R = \phi_R^{[T_1]} - \phi_R^{[R_2]} + \phi_R^{[T_2]} - \phi_R^{[R_1]}$ needs to be constant over the set of frequencies $f_R + c\Delta f$. This requirement can be achieved by an appropriate transceiver architecture, i.e. a common local oscillator (LO) for transmitter and receiver $\phi_R^{[T_x]} = \phi_R^{[R_x]}$ yielding $\Delta\phi_R = 0$ or numerical intermediate frequency mixing $c\Delta f$ for both transmitter T_x and receiver R_x and continuous operation of the radio frequency LO at f_R leading to $\Delta\phi_R = const.$ [8].
- The combination of (9) and (10) in (11) to eliminate unknown time reference t_0, results in the square of the channel transfer function $H^2(f)$. Due to the 2π ambiguity of phase measurements, recovering H_c with the square-root operation introduces a 1π ambiguity per channel. Techniques to estimate the resulting 2^C-state error function under certain hypothesis and conditions are studied in [11]. For the general case, range estimation can be based on $\tilde{H}_c$, considering supplementary virtual paths in the convoluted channel impulse response $h * h$. For example, in a scenario with $P = 2$, peaks will appear at $2\tau_0$, $\tau_0 + \tau_1$ and $2\tau_1$ with amplitudes α_0^2, $2\alpha_0\alpha_1$ and α_1^2 respectively.
- For such a processing scheme, the 2π ambiguity translates to a range ambiguity $R_{max} = \frac{c_0}{2 \cdot \Delta f}$ and the virtual bandwidth $BW_{virt} = (C - 1)\Delta f$ limits range resolution $\Delta R = \frac{c_0}{2 \cdot BW_{virt}}$ [6].

In the following, four range estimators based on (11) are studied.

Phase Slope. For a single propagation path $(P = 1)$, the estimated channel transfer function (11) is given by $\tilde{H}_c = \alpha_0^2 e^{-j2\pi c\Delta f 2\tau_0}$. Range information can be extracted from the slope of the argument of $\tilde{H}_c$

$$\hat{d}_{slope} = \tau c_0 = -\frac{c_0}{4\pi\Delta f} \frac{\Delta\arg\left(\tilde{H}_c\right)}{\Delta c}. \tag{12}$$

Inverse Discrete Fourier Transform. In multipath scenarios $(P > 1)$, the reconstructed channel transfer function (11) can be converted to an estimate of the convoluted channel impulse response $h * h$ by Inverse Discrete Fourier Transform (IDFT)

$$\tilde{h}_k = \text{IDFT}\left\{\tilde{H}_c\right\}. \tag{13}$$

In order to detect the first/direct path in multipath scenarios with possibly stronger secondary paths, the following *first path detection* scheme is applied. The range estimate $\hat{d}_{\text{IDFT}}$ is taken as the first local maximum above a certain threshold γ relative to and in a certain range R_{first} before the global maximum in the estimated channel impulse response $\tilde{h}_k$. Sufficient zero padding ensures fine time and hence range granularity. However, range resolution remains bound by the virtual bandwidth BW_{virt}.

MUltiple SIgnal Classification. The estimation of amplitudes α_p and delays τ_p in the reconstructed channel transfer function can be formulated as a spectral estimation problem to which high-resolution techniques such as MUltiple SIgnal Classification (MUSIC) can be applied [13]. Therefore (11) with (3) is expanded and rewritten as

$$\tilde{H}_c = \sum_{p=0}^{\tilde{P}-1} \tilde{\alpha}_p e^{-j2\pi(f_R + c\Delta f)\tilde{\tau}_p}, \tag{14}$$

where $\tilde{P}$, $\tilde{\tau}_p$ and $\tilde{\alpha}_p$ account for the notation after expansion. Analog to [13], defining an imaging vector

$$\phi^C(\tilde{\tau}_p) = \left[1\, e^{-j2\pi\Delta f\tilde{\tau}_p} \, \ldots \, e^{-j2\pi(C-1)\Delta f\tilde{\tau}_p}\right]^{\text{T}} \tag{15}$$

and

$$\boldsymbol{\Phi}^C = \left[\phi^C(\tilde{\tau}_0)\, \phi^C(\tilde{\tau}_1) \, \ldots \, \phi^C(\tilde{\tau}_{\tilde{P}-1})\right], \tag{16}$$

allows reformulating (14) in matrix form as

$$\tilde{\boldsymbol{H}} = \left[\tilde{H}_0 \, \ldots \, \tilde{H}_{C-1}\right]^{\text{T}} = \boldsymbol{\Phi}^C \tilde{\boldsymbol{\alpha}} + \tilde{\boldsymbol{N}}, \tag{17}$$

with amplitude $\tilde{\boldsymbol{\alpha}} = \left[\tilde{\alpha}_0\, \tilde{\alpha}_1 \, \ldots \, \tilde{\alpha}_{\tilde{P}-1}\right]^{\text{T}}$ and noise $\tilde{\boldsymbol{N}}$.

The construction of the Hankel matrix

$$\boldsymbol{K} = \begin{pmatrix} \tilde{H}_0 & \tilde{H}_1 & \cdots & \tilde{H}_{C-L-1} \\ \tilde{H}_1 & \tilde{H}_2 & \cdots & \tilde{H}_{C-L} \\ \vdots & \vdots & \vdots & \vdots \\ \tilde{H}_L & \tilde{H}_{L+1} & \cdots & \tilde{H}_{C-1} \end{pmatrix}, \tag{18}$$

with $1 \leq L < C$, allows the application of MUSIC to a single snapshot of the channel transfer function estimates $\tilde{H}$ [13]. The singular value decomposition of the Hankel matrix (18) is given by

$$\boldsymbol{K} = [\boldsymbol{U}_S \, \boldsymbol{U}_N] \operatorname{diag}\left(\lambda_0 \, \lambda_1 \, \ldots \, \lambda_{\tilde{P}-1} \, 0 \, \ldots \, 0\right) [\boldsymbol{V}_S \, \boldsymbol{V}_N]^*, \tag{19}$$

with singular values $\lambda_0 \geq \lambda_1 \geq \ldots \geq \lambda_{\tilde{P}-1} \geq 0$ and complex conjugate $(\cdot)^*$.

The set of delays $\tilde{\tau}_p$ can be obtained as the peaks in the imaging function Y, defined by the orthogonal projection of the imaging vector ϕ^L to the noise subspace $\boldsymbol{U}_N$

$$Y(\tau) = \left\| \phi^L(\tau) \right\|_2 \Big/ \left\| \boldsymbol{U}_N^* \phi^L(\tau) \right\|_2 \tag{20}$$

The range estimate $\hat{d}_{\mathrm{MUSIC}}$ is obtained by applying the *first path detection* to (20).

In contrast to the IDFT approach, delay estimation by the MUSIC algorithm can achieve arbitrary high resolution for sufficiently low noise [13].

Deep Neural Network Based Regression Approach. The problem considered so far can be viewed as determining a function $f^\theta : \mathbb{C}^C \to \mathbb{R}$ parameterized by θ, which computes the range based on channel measurements. To find this function, thus the parameters θ, a Deep Neural Network (DNN) based regression algorithm is conducted on an expert database $\mathcal{D} = \left\{ (\{\tilde{H}_{c,i}\}_{c=0}^{C-1}; d_i) \right\}_{i=1}^N$, where $N = |\mathcal{D}|$ is the database size. Each entry i in the database consists of a collection of channel measurements $\{\tilde{H}_{c,i}\}_{c=0}^{C-1}$ and the associated true range value d_i. During the training, the network parameters θ are updated via a gradient descend algorithm to minimize the loss function

$$\mathcal{L}(\theta) = \mathbb{E}\left[(f^\theta(\{\tilde{H}_{c,i}\}_{c=0}^{C-1}) - d_i)^2 \right]. \tag{21}$$

During the exploitation phase, the range estimation $\hat{d}_{DNN}$ for a given channel measurement $\{\tilde{H}_c\}_{c=0}^{C-1}$ is obtained from the network parameter estimates $\hat{\theta}$ as $\hat{d}_{DNN} = f^{\hat{\theta}}(\{\tilde{H}_c\}_{c=0}^{C-1})$.

3   Numerical Simulations

3.1   Parametric Estimators

Numerical simulations are conducted to evaluate the performance of the presented range estimators on synthetic data. Simulations consider a two-path propagation scenario with a direct and a multipath component, where $\beta = \alpha_1/\alpha_0$ and $\Delta\tau = \tau_1 - \tau_0$. The reconstructed channel transfer function estimates $\tilde{H}_c$ are obtained by (11) with $C = 16$ and $\Delta f = 200\,\mathrm{kHz}$ leading to $BW_{\mathrm{virt}} = 3\,\mathrm{MHz}$, $\Delta R = 50\,\mathrm{m}$ and $R_{\mathrm{max}} = 750\,\mathrm{m}$. AWGN of variance σ^2 is added to each channel estimate $H_c^{[T,R]}$. *First path detection* is performed on (13) and (20) with the optimized parameters $R_{\text{first}} = 300\,\text{m}$ and $\gamma = -7\,\text{dB}$. MUSIC requires $P_{\text{MUSIC}} = \tilde{P} = 3$ for two-path propagation. For each channel parameter configuration $(E_s/N_0, \beta, \Delta\tau)$, 5000 Monte Carlo simulations are performed.

Figure 2a, Fig. 2c and Fig. 2e compare the maximum ranging error in 90% of the cases for the IDFT $\hat{d}_{\text{IDFT}}$ and MUSIC $\hat{d}_{\text{MUSIC}}$ estimator as function of the path delay difference $\Delta\tau$. For a path delay difference approximately equal to the multi-channel range resolution ($\Delta\tau \approx \Delta R$), a maximum error is observed as overlapping main lobes form a larger lobe. The precision gain of MUSIC over the IDFT technique is most pronounced for path delay differences $\Delta\tau > 100\,\text{m}$ as MUSIC resolves multiple propagation paths and precision asymptotically attains single path precision. Range estimation degrades as the path amplitude ratio β increases, which makes the detection of a weak first path difficult. For $\beta = 10\,\text{dB}$, *first path detection* does no longer find the first path below the threshold γ.

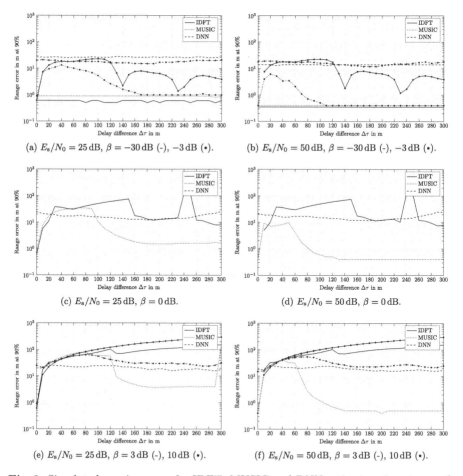

(a) $E_s/N_0 = 25\,\text{dB}$, $\beta = -30\,\text{dB}$ (-), $-3\,\text{dB}$ (•).

(b) $E_s/N_0 = 50\,\text{dB}$, $\beta = -30\,\text{dB}$ (-), $-3\,\text{dB}$ (•).

(c) $E_s/N_0 = 25\,\text{dB}$, $\beta = 0\,\text{dB}$.

(d) $E_s/N_0 = 50\,\text{dB}$, $\beta = 0\,\text{dB}$.

(e) $E_s/N_0 = 25\,\text{dB}$, $\beta = 3\,\text{dB}$ (-), $10\,\text{dB}$ (•).

(f) $E_s/N_0 = 50\,\text{dB}$, $\beta = 3\,\text{dB}$ (-), $10\,\text{dB}$ (•).

Fig. 2. Simulated ranging error for IDFT, MUSIC and DNN estimators in a two-path propagation scenario.

Ranging errors then grow linearly with delay difference $\Delta\tau$. These findings are confirmed at high E_s/N_0, where MUSIC clearly outperforms the IDFT approach as depicted in Fig. 2b, Fig. 2d and Fig. 2f.

3.2   DNN Based Range Estimation Algorithm

The database $\mathcal{D}$ is generated via Monte Carlo simulations. Table 1 summarizes the parameters used during simulations. Figure 3 shows the DNN architecture, which comprises three hidden layers of 128 units. Each neuron has a Rectifier Linear Unit activation (ReLU) except the final layer. For a practical representation, each channel measurement is unpacked into real and imaginary part, thus, the input of the DNN is of size $2C$. During the learning phase, a minibatch of 32 samples is randomly taken every iteration from the database on which the gradient descend is performed with an empirical step size or learning rate $\alpha = 0.001$. Figure 2 shows that the resulting DNN algorithm has comparable performance with respect to both IDFT and MUSIC algorithms. While for $\beta < 0\,\mathrm{dB}$ it slightly performs less than the parametric methods, for $\beta \geq 0\,\mathrm{dB}$, the precision of the DNN algorithm takes over when $\Delta\tau < 100\,\mathrm{m}$ and at low E_s/N_0 but still remains lower than MUSIC technique when $\Delta\tau$ increases. In addition, it is worth to highlight that the proposed DNN algorithm is less sensitive to the variation of the parameters β, $\Delta\tau$, and E_s/N_0. In fact, the range estimation error almost remains constant ($\approx 25\,\mathrm{m}$) when these parameters vary, demonstrating therefore the robustness of the DNN algorithm.

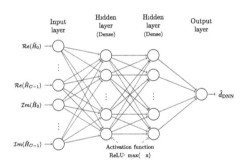

Fig. 3. Illustration of the DNN architecture for multi-channel range estimation.

Table 1. DNN training data generation parameters.

Parameter	Symbol	Value
Path amplitude ratio	$\beta = \alpha_1/\alpha_0$	$-30:5:$ $-5, -3, -1,$ $0, 1, 3, 5, 10\,\mathrm{dB}$
Path delay difference	$\Delta\tau = \tau_1 - \tau_0$	$0:10:200\,\mathrm{m}$
Signal-to-noise ratio	E_s/N_0	$-30:5:$ $50\,\mathrm{dB}$
Monte Carlo runs		1500

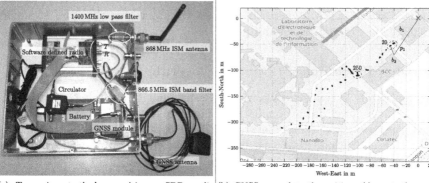

(a) Transceiver testbed comprising a SDR, radio frequency components, a GNSS module and a power supply.

(b) GNSS ground truth positions (·) w.r.t. the roof testbed (×) at $(0,0)$.

Fig. 4. Multi-channel ranging field trials.

4   Application to Field Trials

4.1   LPWA Ranging Transceiver Testbed

The transceiver testbed [8] illustrated in Fig. 4a comprises a Software Defined Radio[2] (SDR), radio frequency components and a Global Navigation Satellite System (GNSS) module.

Digital intermediate frequency up-/down-mixing stages are implemented into the SDR, allowing to coherently process a 10 MHz bandwidth by sequentially selecting 1 MHz channels. Inphase/quadrature (IQ) samples are stored for offline processing by a dedicated software.

Two transceiver testbeds perform a multi-channel two-way ranging protocol according to Fig. 1, exchanging a Binary Phase Shift Keying (BPSK) preamble of chip rate $R_c \equiv BW_{\text{sym}} = 10\,\text{kHz}$ following a Gold code of chip length $N_c = 256$ in the 868 MHz Industrial Scientific Medical (ISM) band. From $T_g = 3\,\text{ms}$ and $T_p = N_c/BW_{\text{sym}} = 25.6\,\text{ms}$ follows a total two-way multi-channel duration $T_{\text{MC}} = 2C\,(T_p + T_g) = 915.2\,\text{ms}$.

The transceiver testbed comprises a uBlox GNSS module[3] for the purpose of providing a ranging ground truth reference. In the following field trials, GNSS position estimates show a standard deviation $\sigma_{\text{GNSS}} < 1\,\text{m}$. Therefore, GNSS position errors are neglected and GNSS estimates are directly used to establish LPWA narrowband ranging errors.

---

[2] *Analog Devices AD9361* 2 × 2TRX radio front end and a *Zynq-045 Xilinx System on chip FPGA* with integrated dual Cortex-A9 ARM processor.

[3] uBlox C94-M8P application board.

4.2   Outdoor Field Trial

Outdoor field trials are performed between a node on the roof of a four-story building and a mobile node on the ground. GNSS reference positions for the measurements are depicted in Fig. 4b. A total of 900 valid measurements in stationary conditions is processed for range estimation. As indicated on the map in Fig. 4b, multipath propagation is possible due to surrounding buildings on the semi-urban industrial site.

4.3   Ranging Results and Discussion

In contrast to simulation, where $H_c^{[T,R]}$ are generated directly, field trial channel estimates $H_c^{[T,R]}$ for the construction of (11) are obtained by cross correlation of the received signals $s_{BB,c}^{[T,R]}$ at the channel frequencies $c\Delta f$ with the transmit waveform s_0 $H_c^{[T,R]}$ [8].

Figure 5 shows the ranging errors for the different range estimation strategies. IDFT ranging precision in Fig. 5a is below 30 m, however for certain positions large biases (>100 m) are observed. Processing the field trial measurements with the MUSIC algorithm in Fig. 5b does not decrease these biases, despite the high-resolution capacity. Possible causes are insufficient signal-to-noise ratio E_s/N_0 (compare Fig. 2) or the mismatch between the hypothesis of a two-path channel ($\tilde{P} = 3$) and the real propagation channel.

Regarding the DNN approach, when trained on synthetic data and then applied on field trial data, the DNN algorithm (S-DNN) slightly performs less than the two parametric estimators (Fig. 5c). This is again because the hypothesis of a two-path channel is no longer valid in real environment, leading to channel measurements that were not seen in the synthetic database. The Cumulative Distribution Functions (CDF) in Fig. 6 summarize the ranging accuracy of the performed field trial.

In order to overcome the weakness of an incomplete database, training can be performed on a portion of the real field trial data (Fig. 5d). Figure 7 shows the CDF (R-DNN) when trained on 2/3 and applied to the remaining 1/3 of the real field trial data in comparison to the performances when the synthetically-only trained DNN is applied to the same 1/3 (S-DNN).

Furthermore, better results can be obtained by taking advantage of synthetic data, by using these data to pre-train the DNN and then refine it with some field trial data (M-DNN) [19], thus reducing the need of an extensive geo-referenced channel measurements database. Figure 5e and Fig. 7 show the results for this approach. The resulting M-DNN has better performance than both S-DNN and R-DNN, therefore demonstrating the benefit of taking into consideration both real and synthetic data for a robust algorithm.

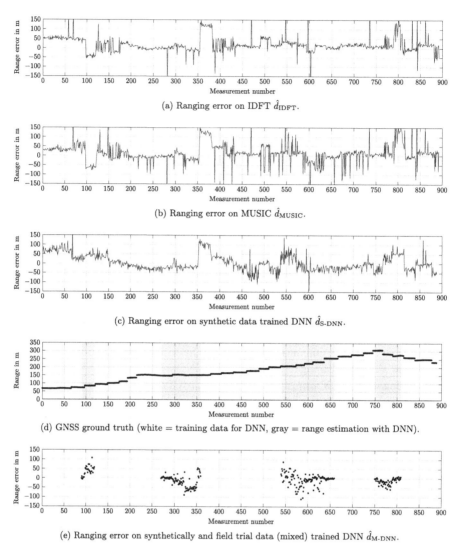

(a) Ranging error on IDFT $\hat{d}_{\text{IDFT}}$.

(b) Ranging error on MUSIC $\hat{d}_{\text{MUSIC}}$.

(c) Ranging error on synthetic data trained DNN $\hat{d}_{\text{S-DNN}}$.

(d) GNSS ground truth (white = training data for DNN, gray = range estimation with DNN).

(e) Ranging error on synthetically and field trial data (mixed) trained DNN $\hat{d}_{\text{M-DNN}}$.

Fig. 5. Multi-channel narrowband ranging errors for the 900 field trial measurements with 10 kHz bandwidth signals.

In conclusion, parametric estimators and synthetic data trained deep learning achieve a ranging error below 60 m to 80 m in 90% of the cases and mixed data trained deep learning attains 45 m.

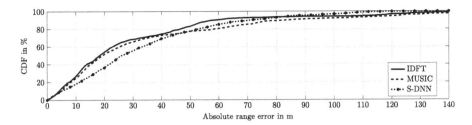

Fig. 6. Cumulative Distribution Functions (CDF) for field trial multi-channel narrowband range estimation with 10 kHz bandwidth signals on all 900 measurements.

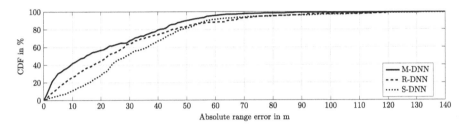

Fig. 7. CDFs for field trial multi-channel narrowband range estimation with 10 kHz bandwidth signals on 1/3 of the measurements (Fig. 5d, gray).

5   Conclusion

Multi-channel narrowband LPWA ranging accuracy has been evaluated by numerical simulations in a two-path propagation scenario. Parametric range estimation through MUSIC outperforms the state of the art Inverse Discrete Fourier Transform approach, due to its high-resolution property. For sufficient large E_s/N_0, MUSIC achieves more accurate range estimates for close multipath than the IDFT technique in numerical simulations. Applied to real outdoor field trials, MUSIC and IDFT estimators show comparable performances (60 to 80 m in 90% of the cases).

The deep learning based range estimation algorithm shows high robustness to multipath, with 25 m error in simulation and 45 m in 90% of the cases on the field trial data, when trained on mixed data. Multi-channel measurements provide improved temporal resolution through sequentially increased bandwidth. Combined with deep learning techniques they are potential enablers for precise LPWA localization due to appropriated signal processing, especially for unknown and dense multipath propagation scenarios. Yet, generalization of these findings to other scenarios remains open.

Further work may consider training deep neural networks assuming more complex channel models. Extensive field trials will provide a comprehensive database for training with mixed synthetic and real data.

Future studies may investigate the combination of multi-channel ranging with beamforming strategies to mitigate multipath as well as Time-Difference-of-Arrival (TDoA) like approaches to address the complexity of the two-way signaling scheme and to ensure compatibility with the star topology of LPWA networks.

References

1. Raza, U., Kulkarni, P., Sooriyabandara, M.: Low power wide area networks: an overview. IEEE Commun. Surv. Tutor. **19**(2), 855–873 (2017). https://doi.org/10.1109/COMST.2017.2652320
2. Razavi, S.M., et al.: Positioning in cellular networks: past, present, future. In: 2018 IEEE Wireless Communications and Networking Conference (WCNC), pp. 1–6, April 2018. https://doi.org/10.1109/WCNC.2018.8377447
3. Link Labs: Lora Localization (2016). https://www.link-labs.com/blog/lora-localization
4. Gotthard, P., Jankech, T.: Low-cost car park localization using RSSI in supervised LoRa mesh networks. In: 2018 15th Workshop on Positioning, Navigation and Communications (WPNC), pp. 1–6, October 2018. https://doi.org/10.1109/WPNC.2018.8555792
5. Aernouts, M., Bellekens, B., Berkvens, R., Weyn, M.: A comparison of signal strength localization methods with sigfox. In: 2018 15th Workshop on Positioning, Navigation and Communications (WPNC), pp. 1–6, October 2018. https://doi.org/10.1109/WPNC.2018.8555743
6. Skolnik, M.: Radar Handbook, 2nd edn. McGrawHill, New York City (1990)
7. Podevijn, N., et al.: TDoA-based outdoor positioning with tracking algorithm in a public LoRa network. Wirel. Commun. Mob. Comput. **2018**, 9 (2018)
8. Wolf, F., Dore, J.B., Popon, X., de Rivaz, S., Dehmas, F., Cances, J.P.: Coherent multi-channel ranging for narrowband LPWAN: simulation and experimentation results. In: 15th Workshop on Positioning, Navigation and Communications (WPNC), pp. 1–6, October 2018
9. Vasisht, D., Kumar, S., Katabi, D.: Decimeter-level localization with a single WiFi access point (2016)
10. Pichler, M., Schwarzer, S., Stelzer, A., Vossiek, M.: Multi-channel distance measurement with IEEE 802.15.4 (ZigBee) devices. IEEE J. Sel. Top. Sig. Process. **3**(5), 845–859 (2009). https://doi.org/10.1109/JSTSP.2009.2030935
11. Schwarzer, S.: Entwicklung eines industriellen Funkortungssystems basierend auf der kohaerenten Kombination von Kommunikationssignalen mit IEEE-802.15.4-Geraeten. Ph.D. thesis, Technischen Universitaet Clausthal (2011)
12. Povalac, A., Sebesta, J.: Phase difference of arrival distance estimation for RFID tags in frequency domain. In: 2011 IEEE International Conference on RFID-Technologies and Applications. pp. 188–193, September 2011. https://doi.org/10.1109/RFID-TA.2011.6068636
13. Liao, W., Fannjiang, A.: MUSIC for single-snapshot spectral estimation: stability and super-resolution. CoRR abs/1404.1484 (2014)
14. Li, X., Pahlavan, K.: Super-resolution TOA estimation with diversity for indoor geolocation. IEEE Trans. Wirel. Commun. **3**(1), 224–234 (2004). https://doi.org/10.1109/TWC.2003.819035
15. Chehri, A., Fortier, P., Tardif, P.: On the TOA estimation for UWB ranging in complex confined area. In: 2007 International Symposium on Signals, Systems and Electronics, pp. 533–536, July 2007. https://doi.org/10.1109/ISSSE.2007.4294530

16. Mao, Q., Hu, F., Hao, Q.: Deep learning for intelligent wireless networks: a comprehensive survey. IEEE Commun. Surv. Tutor. **20**(4), 2595–2621 (Fourthquarter 2018). https://doi.org/10.1109/COMST.2018.2846401
17. Bialer, O., Garnett, N., Levi, D.: A deep neural network approach for time-of-arrival estimation in multipath channels. In: 2018 IEEE International Conference on Acoustics, Speech and Signal Processing (ICASSP), pp. 2936–2940, April 2018. https://doi.org/10.1109/ICASSP.2018.8461301
18. Niitsoo, A., Edelhäußer, T., Eberlein, E., Hadaschik, N., Mutschler, C.: A deep learning approach to position estimation from channel impulse responses. Sensors **19**(5), 1064 (2019)
19. Zappone, A., Di Renzo, M., Debbah, M., Lam, T.T., Qian, X.: Model-aided wireless artificial intelligence: embedding expert knowledge in deep neural networks towards wireless systems optimization. arXiv preprint arXiv:1808.01672 (2018)

Iterative Decoding for SCMA Systems Using Log-MPA with Feedback LDPC Decoding

Bilal Ghani[1]([✉]), Frederic Launay[1], Jean Pierre Cances[2], Clency Perrine[3], and Yannis Pousset[3]

[1] LIAS (EA 6315), Université de Poitiers, 2 rue Pierre Brousse, 86073 Poitiers Cedex 9, France
{bilal.ghani,frederic.launay}@univ-poitiers.fr
[2] XLIM (UMR CNRS 7252), Université de Limoges, 123 avenue Albert Thomas, 87060 Limoges Cedex, France
cances@ensil.unilim.fr
[3] XLIM (UMR CNRS 7252), Université de Poitiers, 11 bd Marie et Pierre Curie, 86073 Poitiers Cedex 9, France
{clency.perrine,yannis.pousset}@univ-poitiers.fr

Abstract. Sparse Code Multiple Access (SCMA) is a technique to improve spectral efficiency by sharing the same spectral resources among different users. As a result, the receiver has to cope with excess interference. A channel encoder is used to aid receiver to correct the data stream. In this paper, we discuss the use of an LDPC encoder for channel coding on each user's traffic, followed by an SCMA encoder. SCMA encoder is a potential candidate for fifth generation (5G) mobile communication and different codebooks have already been proposed. In this paper we choose specific codewords extracted from permutation of QPSK modulation and rotation. The receiver is based on Log Domain Message Passing Algorithm (Log-MPA) for SCMA. The novelty of our paper lies in the receiver scheme with exchange of extrinsic information between SCMA decoder and LDPC decoder. Feedback from LDPC decoder provides extrinsic information to SCMA decoder, thus enabling creation of a super-graph between LDPC and SCMA decoder. Numerical results depict performance gain in terms of Bit Error Rate (BER) and complexity over conventional LDPC coded SCMA systems.

Keywords: Sparse code multiple access · Log-massage passing algorithm · 5G · IoT · LDPC · LLR · Iterative decoding

1 Introduction

The Internet of Things (IoT) is a network of scattered physical objects (things) which sense external environment and communicate their internal state to a server application. The pace of innovation has generated requirements for billions

© Springer Nature Switzerland AG 2020
O. Habachi et al. (Eds.): UNet 2019, LNCS 12293, pp. 18–31, 2020.
https://doi.org/10.1007/978-3-030-58008-7_2

of devices, primarily wireless device and until one million wireless devices per km^2 are expected to be handled by a cellular gateway.

An auspicious non-orthogonal multiple-access technique which can simultaneously support massive connections is SCMA, making it a competitive candidate for 5G communication. The performance of SCMA is mainly dependent by the codebooks [1] and the receiver algorithm. SCMA is a codebook division multiple access scheme, which is characterized by the sparsity of code. Appertaining to multidimensional constellations, codebooks are constructed, which transcend conventional spread code based schemes [2]. The sparsity is benefit for the receiver complexity, and MPA is applied to achieve near optimal performance. Even if codes are sparse, multiple users share the same frequency resources with different codes which bring collisions between users (MAI: Multiple Address Interference).

The complexity of the receiver algorithm depends on the choice of codebook. MPA (Message Passing Algorithm) presents good performance with lower complexity [3] compared to optimum MAP (Maximum a Posteriori). Nevertheless, computation complexity of SCMA is still relatively high and optimization can be achieved by exploring the lattice structure of SCMA codewords [4]. Moreover, many exponential computations are involved in MPA to calculate probabilities and extrinsic information of received signal. [5] proposes a simplified Log-domain MPA to compute extrinsic information in log domain. [6] proposes bit error rate (BER) performance and complexity of Log-MPA is improved as compared to MPA. To reduce the complexity of receiver, we have chosen an optimized SCMA codebook [7]. Log-MPA is used in SCMA decoding part. According to [8], the latest 3GPP report, LDPC and Polar are the two selected codes in 5G standard. Using LDPC codes as channel coding, iterative detection and decoding has been done for SCMA systems in [9].

The goal of this paper is to develop a combination of Log-MPA decoding for SCMA with LDPC decoding. The novelty in this paper is the addition of feedback from LDPC decoder to Log-MPA decoder enabling the receiver to exchange extrinsic information between the two blocks. Moreover complexity of [9] is further reduced by exploiting random generation of parity check matrix, extra inter-leavers in [9] are removed.

The rest of the paper is organized as follows. Section 2 briefly discusses the design of SCMA users codebook, then during transmission part, LDPC encoder and SCMA encoder is discussed. Receiver part discusses the iterative log-message passing decode followed by LDPC decode after that feedback from LDPC to Log-MPA is discussed. Simulation results are disused in Sect. 3 to illustrate performance gain. Finally, Sect. 4 concludes the paper.

2 System Model

We consider the transmission of LDPC coded SCMA system with J users transmitting over K resource blocks. Input data bits from each user $j = 1, ..., J$ are $d_{m_o}^j | m_o = 1, ..., m$. These data bits are encoded by LDPC encoder to

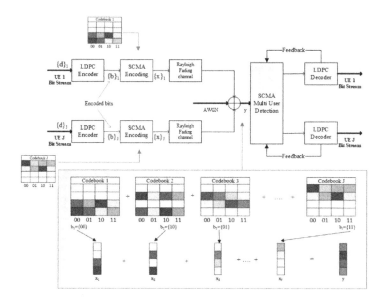

Fig. 1. LDPC coded SCMA system with feedback

$b_{n_o}^j | n_o = 1, ..., n$, with a code rate $R = m/n$. Every $log_2(M)$ encoded bits are group together and an SCMA encoder maps them to K-dimensional complex symbol of size M. These complex symbols are signaled through Rayleigh fading channel with additive white Gaussian noise (AWGN), as shown in Fig. 1. We have considered six users i.e. $j = 1, ..., 6$ transmitting data over four resource blocks (RBs) so $K = 4$. Code rate for LDPC encoder is $R = 1/2$.

2.1 Design of Codebooks

Codebooks are designed in accordance to factor graph representation mentioned in [1]. We have taken four RBs, i.e. $K = 4$ while $J = 6$ is the number of users. Overloading factor $\lambda = J/K = 1.5$. Quadrature Phase Shift Keying (QPSK) constellation has been used to generate codewords from which codebooks for each user are generated. N is chosen to be small enough in comparison to K to maintain sparsity. $N = 2$ is chosen so two dimensional constellation points can be mapped on $K > 2$ resources to construct SCMA codewords.

$$F = \begin{bmatrix} 1 & 1 & 1 & 0 & 0 & 0 \\ 1 & 0 & 0 & 1 & 1 & 0 \\ 0 & 1 & 0 & 1 & 0 & 1 \\ 0 & 0 & 1 & 0 & 1 & 1 \end{bmatrix} \qquad (1)$$

F is the graph model which represents the RB used for each user. If $F_{ij} = 1$ then user j transmits a codeword to the resource i. The number resource occupied by a layer depends on the mapping matrix V_j and this set is determined by inserting

$K - N = 2$ all-zero row vectors within the rows of I_N based on the factor graph representation F.

$$F = (f_1, f_2, ...f_J) where f_J = diag(V_j V_j^T) \tag{2}$$

Once we have the mapping matrix V_j for each layer j along with the codeword from mother constellation (MC), codebooks for user can be calculated by

$$C_j = V_j(\Delta_j)C_{MC} \tag{3}$$

Typical operators Δ_j defined in [10] are complex conjugate, vector permutation and phase operator. The factor graph in Fig. 2 shows the connection between users and resource blocks. $J = 6$ users connected to $K = 4$ resource blocks.

Fig. 2. Factor graph representation of SCMA system

Each RB is connected to 3 users i.e. $d_f = 3$. The complexity of MPA detection is proportional to M^{d_f}

2.2 LDPC Encoder

LDPC encoder were originally invented by Gallager [11] in the early 1960's. LDPC encoding is implemented as a linear operation from a generator binary matrix: Let's $x = x_1,...x_m$ a m-dimensional message node, $x = \{0,1\}^m$, the encoded message is the image of G : $z = mod(x * G, 2)$ where G is a binary matrix $m \times n$.

z is the encoded message also name codeword of size n. The efficiency of LDPC encoder is $R = m/n$ with $n = k + m$ and k is the number of check constraints added to signal. Efficiency of LDPC is to detect errors of the received noisy signal since the received signal should satisfy the k parity check constraints. Let's $c = c_1, ..., c_k$ the k-dimensional check node, c_j check the parity of a combination of l message bits ($1 \leq i_1 < ... < i_l \leq m$) of the m-dimensional message x : $x_{i_1} \oplus x_{i_2} \oplus ... \oplus x_{i_l} \oplus c_j = 0$.

$m \times n$ binary matrix is the parity check matrix denoted by H. Equations are represented by rows while digits in the codeword are represented by columns. As decoding complexity and a minimum distance increase linearly with the code length, it the sparseness of H which guarantees both. An LDPC code parity-check matrix is called regular if each check bit c_i, $1 \leq i \leq k$ is a constraint of a fixed number of message bits, and each digit in the codeword used in a check function is the same.

MacKay–Neal [12] proposed an other construction of LDPC codes. In proposed structure of parity check matrix, columns are added one by one from left to right. The weight of each column is carefully chosen and non-zero entries in each column is inserted randomly from unfilled rows. If at any instant there are certain empty row positions, more than number of columns to be added then we cannot have exact row degree distribution of H. In this case, the process of adding columns can be restarted or it can be shifted back to insertion of few columns, till correct row degree is obtained. In this article, we use a sparse MacKay–Neal parity check matrix H with $R = 1/2$

2.3   SCMA Encoder

In an SCMA system when J users are transmitting over K physical layers, an SCMA encoder maps the incoming coded binary data bits $b_{n_o}^j$ to K-dimensional complex constellations. After the binary data bits $d_{m_o}^j$ are encoded to $b_{n_o}^j$ coded binary data bits, they are group together to $log_2(M)$ bits and mapping is done as $f : \mathbb{B}^{log_2(M)} \rightarrow \chi$, $x = f(b)$ where $\chi \subset \mathbb{C}^K$. The K-dimensional complex codeword x is a sparse vector with $N < K$ non-zero entries. Each layer has its own codebooks and to generates its own desired codewords depending on it $log_2(M)$ group incoming data. If c is N^{th} dimensional complex constellation point within the constellation set, $c \subset \mathbb{C}^N$ then $log_2(M)$ grouped binary encoded data is mapped on c as $\mathbb{B}^{log_2(M)} \rightarrow c$. As a whole SCMA encoder uses a binary mapping V mentioned in (2) where $V \in \mathbb{B}^{K \times N}$ and having $K - N$ all zero rows maps N dimensional constellation point to a K dimensional codeword. Figure 1 shows the multiplexing of all chosen codewords from respective j users are added together. In fact each of the transmitted codeword is multiplied by the channel coefficient prior to their addition. The received signal after multiplexing can be expressed as

$$y = \sum_{j=1}^{J} diag(h_j)x_j + n \tag{4}$$

where $x_j = (x_{1j}, ..., x_{Kj})^T$ is the SCMA codeword of layer j, $h_j = (h_{1j}, ..., h_{Kj})^T$ is the channel vector of layer j and n is the white additive noise.

2.4   Iterative SCMA Decoding with Log-MPA

Exploiting the sparsity of SCMA codewords, near optimal performance can be achieved by message passing algorithm on factor graph. MPA has lower complexity than Maximum a posterior probability (MAP) detection but still due to numerous exponential operations included in MPA for LLR computation, increases its implementation complexity. This problem with exponential computation in MPA can be catered by using log domain computations. In [13],

Log-MAP decoding algorithm are adopted as an inspiration for implementing MPA in log domain to further decrease the implementation complexity. In presented model, the possible codewords for each of connected users is 4 and there are 3 connected users at each resource block. Thus making total output of possible combination of codewords at each resource block be 64. So, the initial step of Log-MPA detection is calculation of all possible distances between each possible combination of codewords and actual received signal

$$\Psi_K(y, m, h) = abs(y_K - \sum_{l \subset \zeta, m_u = 1:K} h_{l,m_u} x_{l,m_u}) \tag{5}$$

where ζ is the set of users connected to resource K, from (1) for $K = 4$ i.e. 4-th RB, (5) becomes

$$\Psi_4(y_4, m_3, m_5, m_6, h_3, h_5, h_6) = abs(y_4 - (h_3 x_3(m_3) + h_5 x_5(m_5) + h_6 x_6(m_6)))$$

for $m_3, m_5, m_6 = 1, 2, 3, 4$ since UE3, UE5, UE6 are connected to RB4. These Euclidean distances can be expressed as probabilities assuming we have perfect channel estimation and Gaussian noise.

$$\Upsilon_K = exp(\frac{-\Psi_K^2}{2\sigma^2}) \ \forall \ K = 1, .., 4 \tag{6}$$

In log domain calculation, Υ_K will become as

$$\Upsilon'_K = (\frac{-\Psi_K^2}{2\sigma^2}) \ \forall \ K = 1, .., 4 \tag{7}$$

where σ^2 is variance of Gaussian noise. After probabilities of all possible combination of codewords of each of three users at each RB is calculated, iterative message passing algorithm starts exchanging probabilities among users and the RBs according to the factor graph. Before start of iterations, a prior probabilities (ω) of each codewords transmitted by each users is assigned. Initially, each user can transmit any of M codewords, so a prior probabilities is $1/M$ for each j-th user transmitting m_u-th codeword from its codebook.

$$\omega_{j,m_u} = 1/M \tag{8}$$

Each of the user j initial probabilities are assigned as follow to start iterative message passing among users and resource block.

$$\Gamma(v_{l,m_u} \to g) = \omega_{j=l,m_u} \forall \ l \subset \zeta, m_u = 1 : M \tag{9}$$

where ζ is the set of users connected to k-th resource. Log domain transformation for (8) and (9) becomes

$$\omega'_{j,m_u} = log_e(\omega_{j,m_u}) \tag{10}$$

$$\Gamma'(v_{l,m_u} \to g) = \omega'_{j=l,m_u} \forall \ l \subset \zeta, m_u = 1 : M \tag{11}$$

Now message passing among users and resource block starts iteratively. First of all each resource block g_K passes information to its neighbouring user v_j. For example v_1, v_2 and v_3 are connect to g_1, so the information sent to v_1 from g_1 is actually extrinsic information to v_1 from v_2 and v_3. This can be observed in the following Fig. 3.

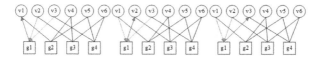

Fig. 3. Extrinsic information passed from g to v

Figure 3 shows the extrinsic information passed to v_1, v_2 and v_3 from g_1. Mathematically, this message passing is expressed as for all $m_u = 1 : M, k = 1 : K$

$$\Gamma(g_{k,m_u} \to v_{l,m_u}) = \sum_{i=1}^{M}\sum_{j=1}^{M} \Upsilon_k(m_u, i, j) \times \Gamma(v_{l_1',i} \to g_{k,m_u}) \times \Gamma(v_{l_2',j} \to g_{k,m_u})$$
(12)

where $l \subset \zeta$ and $l_1', l_2' \subset \zeta - \{l\}$ with $l_1' \neq l_2'$ where ζ is the set of users connected to k-th resource. (12) represents the message passing for MPA decoding, since it is using (6) and (9). For Log-MPA we convert (12) to log domain, using (7) and (11)

$$\Upsilon_k(m_u, i, j) \times \Gamma(v_{l_1',i} \to g_{k,m_u}) \times \Gamma(v_{l_2',j} \to g_{k,m_u}) \Rightarrow \Upsilon_k'(m_u, i, j)$$
$$+ \Gamma'(v_{l_1',i} \to g_{k,m_u}) + \Gamma'(v_{l_2',j} \to g_{k,m_u}) \ \forall \ m_u = 1 : M, k = 1 : K$$
(13)

Now we have to compute $\sum_{i=1}^{M}\sum_{j=1}^{M}$ of (12) in log domain. The result of logarithmic addition in (13) is converted to exp and added. Then log of these sum of exponential is computed. For a specific value of m_u and k, $M \times M$ values of (13) are calculated since $i = 1 : M$ and $j = 1 : M$ so log-domain equivalent of (12) becomes as, for all $m_u = 1 : M, k = 1 : K$

$$\Gamma'(g_{k,m_u} \to v_{l,m_u}) = log_e(e_{i,j}^{\Upsilon_k'(m_u,i,j)+\Gamma'(v_{l_1',i} \to g_{k,m_u})+\Gamma'(v_{l_2',j} \to g_{k,m_u})})$$
(14)

for a specific value of m_u and k. The log of sum of $M \times M$ number of exponential can be solved using following expression [14],

$$log_e(e^a + e^b) = max(a, b) + log_e(e^{-|a-b|} + e^{-|b-b|})$$

where $max(a, b) = b$, so generally n exponential sum i.e. $(e^{a_1} + ... + e^{a_n})$ and if $max(a_1 + ... + a_n) = a^*$ then

$$log_e(e^{a_1} + ... + e^{a_n}) = a^* + log_e(e^{-|a_1-a^*|} + e^{-|a_2-a^*|} + ... + e^{-|a_n-a^*|})$$
(15)

Using (15), we can solve log of sum of exponential in (14) to have numerical stability. $\Gamma'(g_{k,m_u} \rightarrow v_{l,m_u})$ is the message passed from resource block to neighbouring nodes. Now the user passes update messages obtained from extrinsic information to its neighboring (connected) resource blocks which can be seen in following Fig. 4.

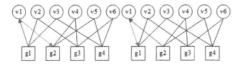

Fig. 4. Extrinsic information passed from v to g

For instance, in the figure above v_1 is connected to g_1 and g_2 and the information sent from v_1 to g_1 is actually extrinsic information v_1 received from g_2. Information from a user to a resource is a normalized guess swap at the sender user node

$$\Gamma(v_{h,j,m_u} \rightarrow g_{h,j,m_u}) = \frac{\omega_{j,m_u} \times \Gamma(g_{h',j,m_u} \rightarrow v_{h,j,m_u})}{\sum_{m_u=1}^{M} \Gamma(g_{h',j,m_u} \rightarrow v_{h,j,m_u})}$$
$$\forall\, j = 1 : J, m_u = 1 : M \tag{16}$$

where $h \subset \varrho$ and $h' \subset \varrho - \{h\}$ and ϱ is set of resource blocks connected to each j-th user. Solving (16) in log domain using values from (10) and (14)

$$\Gamma'(v_{h,j,m_u} \rightarrow g_{h,j,m_u}) = \omega'_{j,m_u} + \Gamma'(g_{h',j,m_u} \rightarrow v_{h,j,m_u})$$
$$- LSE_{m_u=1}^{M}(\Gamma'(g_{h',j,m_u} \rightarrow v_{h,j,m_u}))\,\forall\, j = 1 : J, m_u = 1 : M \tag{17}$$

where $h \subset \varrho$ and $h' \subset \varrho - \{h\}$ and ϱ is set of RBs connected to each j-th user. $LSE_{m_u=1}^{M}$ is Log of Sum of Exponential and is solved using (15). Message passing among users and resources, i.e. (14) and (17) is repeated for three to ten iterations till required decoding is achieved. After iterative message passing among user and RBs is ended, the guess of symbol level LLR for each user is simply a product from all guesses from its neighbouring RBs and *a prior* probability so its simply addition in log domain as follows

$$Q_{j,m_u} = \omega'_{j,m_u} + \Gamma'(g_{h,j,m_u} \rightarrow v_{h,j,m_u})$$
$$+ \Gamma'(g_{h',j,m_u} \rightarrow v_{h,j,m_u})\,\forall\, j = 1 : J, m_u = 1 : M \tag{18}$$

where $h \subset \varrho$ and $h' \subset \varrho - \{h\}$ and ϱ is set of resource blocks connected to each j-th user. At the output of Log-MPA decoder we have symbol level LLR, for i-th instant, $LLR_i^{s_n} = Q_{j,m_u}$ for $s_n, m_u = 1, ..., M$.

2.5 LDPC Decoder

Belief Propagation decoding for LDPC Decode consists of initialization, horizontal step, vertical step and hard decision of input. After the initialization, the remaining steps are computed for k iterations until convergence or maximum number of iterations are achieved. Let n be the variable nodes, m be the check node and μ_n is the initial values provided to LDPC decode i.e. $\mu_n = log_e \frac{Pr(b_n=0)}{Pr(b_n=1)}$. In developed model, μ_n is initially provided by Log-MPA decoder. At k-th iteration α_{nm}^k is the bit level LLR passed from variable node n to check node m, β_{mn}^k is the bit level LLR passed from check node m to variable node n and γ_n^k is *a posteriori* LLR of variable node n.

For i-th instant, symbol level LLR returned by Log-MPA decoder are $LLR_i^{s_n}, s_n = (1, ..., M)$. To feed these values to LDPC decoder we need to convert these symbol level LLR i.e. $LLR_i^{s_n}$ to bit level LLR i.e. $LLR_{i_n}^b, i_n = (1, ..., log_2(M))$. We grouped $log_2(M)$ bits together before mapping them onto their complex codebooks so which have to be un-grouped. Thus for each i-th instant we have $log_2(M)$ values. While using traditional mapping of symbol to QPSK, following is the conversion of symbol level LLR to bit level LLR.

$$LLR_{i_1}^b = log_e \frac{exp(LLR_i^{s_1}) + exp(LLR_i^{s_2})}{exp(LLR_i^{s_3}) + exp(LLR_i^{s_4})} \tag{19}$$

$$LLR_{i_2}^b = log_e \frac{exp(LLR_i^{s_1}) + exp(LLR_i^{s_3})}{exp(LLR_i^{s_2}) + exp(LLR_i^{s_4})} \tag{20}$$

Initialization goes as $\alpha_{nm}^{(0)} = \mu_n = [LLR_{i_1}^b \, LLR_{i_2}^b]$. After this k iterations start for remaining horizontal step, vertical step and hard decision. For each k iteration horizontal step is as

$$\beta_{mn}^{(k)} = 2tanh^{-1}\left(\prod_{n' \in N, n' \neq n} tanh\frac{\alpha_{n'm}^{(k-1)}}{2} \right) \tag{21}$$

vertical step is as

$$\alpha_{nm}^{(k)} = \mu_n + \sum_{m' \in M, m' \neq m} \beta_{m'n}^{(k)} \tag{22}$$

A posteriori LLR of variable node n is thus calculated as

$$\gamma_n^k = \mu_n + \sum_{m' \in M, m' \neq m} \beta_{m'n}^{(k)} \tag{23}$$

Hard decision can be made on the basis of γ_n^k calculated in (23). Implementation complexity of the decoder can be reduced by scheme mentioned in [15]. From (19) and (20), we can calculate likelihood of bit being '0' or '1'. That means we have symbol level LLR from Log-MPA, which is converted to bit level LLR and further converted to likelihoods of being '0' or '1' and then forwarded to LDPC decoder. So in the developed model, at i-th instant, for j-th user, we have M symbol level LLRs that are converted to $log_2(M)$ bit level LLRs that are further converted to $2 \times log_2(M)$ likelihoods of bits being '0' or '1' as shown in Fig. 5

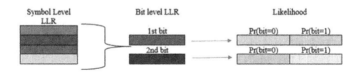

Fig. 5. Symbol Level LLR to bit level LLR

Let $L_{i_1}^{b=0}$ be likelihood for $bit = 0$ and $L_{i_1}^{b=1}$ be likelihood for $bit = 1$. For the next bit at same instant be $L_{i_2}^{b=0}$ and $L_{i_2}^{b=0}$ are likelihoods for $bit = 0$ and $bit = 1$ respectively. These can be calculated using (19) and (20).

$$L_{i_1}^{b=1} = \frac{1}{1 + exp(LLR_{i_1}^b)} \text{ and } L_{i_1}^{b=0} = 1 - L_{i_1}^{b=1} \tag{24}$$

Similarly, $L_{i_2}^{b=0}$ and $L_{i_2}^{b=1}$ are calculated. These values are provided by Log-MPA and are used for initialization for LDPC decoder. Likelihood $L_n^{b=0}$ is assigned to q_{mn}^0 and $L_n^{b=1}$ is assigned to q_{mn}^1. For the horizontal step, we define $\delta q_{mn} = q_{mn}^0 - q_{mn}^1$ and then following is computed for each value of m and n.

$$\delta r_{mn} = \prod_{n' \in N, n' \neq n} \delta q_{mn'} \tag{25}$$

The value of r_{mn} is updated as $r_{mn}^0 = \frac{1}{2(1-\delta r_{mn})}$ and $r_{mn}^1 = \frac{1}{2(1+\delta r_{mn})}$. For the vertical step, for compute the following for each value of m and n

$$q_{mn}^0 = L_n^{b=0} \times (\prod_{m' \in M, m' \neq m} \delta r_{m'n}^0) \text{ and } q_{mn}^1 = L_n^{b=1} \times (\prod_{m' \in M, m' \neq m} \delta r_{m'n}^1)$$

q_{mn} is further updated as,

$$q_{mn}^0 = \frac{q_{mn}^0}{q_{mn}^0 + q_{mn}^1} \text{ and } q_{mn}^1 = \frac{q_{mn}^1}{q_{mn}^0 + q_{mn}^1} \tag{26}$$

A posteriori probabilities are then calculated as, for $b = 0$ and $b = 1$

$$q_n^b = \eta_n \times L_n^b \times (\prod_{m \in M} \delta r_{mn}^b) \tag{27}$$

where η_n is such that $q_n^0 + q_n^1 = 1$ Calculated *a posteriori* probabilities are used for the decoding for k iteration, until the check is validate by parity check matrix or maximum number of iterations are achieved.

2.6   Feedback from LDPC to Log-MPA

If the maximum number of iterations are achieved in LDPC decoder and the check is not validated, means even after maximum iteration, the code has not been able to decoded properly, then feedback is sent to Log-MPA. This serves as extrinsic information to Log-MPA Decoder. This extrinsic information can be obtained by difference of LLR at the bit level at the output of the LDPC decoder and the Log-MPA decoder. Since we have bit level LLR from Log-MPA decoder from Eqs. (19) and (20). So first we calculate bit level LLR at output of LDPC Decoder from Eq. (27) for each user j. For understanding, we denote BL for bit level and SL for symbol level.

$$LLR_n^{j,LDPC_{BL}} = log_e \frac{q_n^{j,0}}{q_n^{j,1}} \tag{28}$$

Now we calculate extrinsic information E_n^j for each user j using (19) and (20) which give $LLR_n^{Log-MPA_{BL}}$ and Eq. (28).

$$E_n^j = LLR_n^{j,LDPC_{BL}} - LLR_n^{j,Log-MPA_{BL}} \tag{29}$$

This extrinsic information from Eq. (29) has to be fed to Log-MPA. As already discussed, Log-MPA decoder deals in symbol level LLR but E_n is the bit level LLR. So first this bit level LLR is converted to symbol level LLR for injection of this extrinsic information to Log-MPA as a feedback. The symbol level LLR for symbol s_i is given in [16]

$$LLR_i = log_e \frac{Pr(y|s = s_i)}{\sum_{k=1,k\neq i}^{M} Pr(y|s = s_k)} \tag{30}$$

From Eq. (29) we calculate the probability for each bit of each user j being '0' or '1' by following

$$Pr_n^{j,b=1} = \frac{1}{1 + exp(E_n^j)} \quad \text{and} \quad Pr_n^{j,b=0} = 1 - Pr_n^{j,b=1}$$

So extrinsic information at symbol level for i-th instance of transmission from j-th user is given by

$$E_i^{j,SL} = log_e \frac{Pr_n^{j,b1} \times Pr_{n+1}^{j,b2}}{\sum_{b1'\neq b1 \,\&\, b2'\neq b2} Pr_n^{j,b1'} \times Pr_{n+1}^{j,b2'}} \tag{31}$$

where $b1, b2, b1'$ and $b2' \in \{0,1\}$. Now we have extrinsic information as symbol level LLR for each i-th instance of transmission from j-th user. This extrinsic information from LDPC decoder can be used as a feedback to Log-MPA two sections, *a prior* probabilities of each codewords transmitted by j-th user and conditional probabilities for given codeword combination of different users at same RB. Referring to (1), exactly three users collide in a one physical resource, that means for each RB we need extrinsic information brought from LDPC Decoder for only those three users colliding on the RB.

$$\xi_K(m) = E_i^j(m) = \sum_{l \subset \zeta, m_u = 1:K} E_{l, m_u} \tag{32}$$

where ζ is the set of users connected to resource K, from (1) for $K = 4$ i.e. 4-th RB, (32) becomes

$$\xi_4(m_3, m_5, m_6) = E_{3, m_3} + E_{5, m_5} + E_{6, m_6} \tag{33}$$

for $m_3, m_5, m_6 = 1, 2, 3, 4$ since UE3, UE5, UE6 are connected to RB4. It returns 64 possible combinations of extrinsic LLR for each resource block, which directly added to conditional probability for given codeword combination. Updated probabilities during initialization process of SCMA decoder (7) becomes

$$\gamma_K^{'\,updated} = (\frac{-\Psi_K^2}{2\sigma^2}) + \xi_K \forall\, K = 1, .., 4 \tag{34}$$

ξ_K is also used to update *a prior* probabilities of each codewords transmitted by j-th user. Initially, during the first round of Log-MPA decode, each user has equal probability of transmitting any of the codewords i.e. $1/M$, from the feedback probability of a codeword being transmitted by j-th user becomes $exp(\xi_K)$. So before start of iteration of message passing (8) is updated as

$$\omega_{j,K} = exp(\xi_{j,K}) \forall\, K = 1, .., 4 | j = 1, ..., 6 \tag{35}$$

Rest of decoding of Log-MPA is followed as already discussed.

3   Simulation Results

In this section, simulation results for Log-MPA decoder with LDPC decoder are discussed, which is a traditional incorporation of channel decoder with SCMA decoder. We also provide the simulation results when feedback is provided from LDPC decoder to Log-MPA decoder.

For both of the models complex symbols are signaled through Rayleigh fading channel with additive white Gaussian noise (AWGN). Number of iteration for Log-Domain message passing is set as $iter_{SCMA} = 10$ and iterations for LDPC decoding is also set as $iter_{LDPC} = 10$. We have used a sparse parity check matrix H with $k = 500$ check nodes and $m = 500$ information bit. The encoded signal is 1000 bits length, i.e. $n = 1000$. The size of H matrix is 500×1000 and the weight of each column is equal to 3 whereas weight of each row is 6 so to make regular parity check matrix H. The code rate $R = 1/2$.

Figure 6 shows the BER of each of j users with Log-MPA decode with and without feedback from LDPC Decoder. It is observed in Fig. 6a BER of 10^{-4} for each user is achieved around 4.5 dB, whereas same BER is achieved at 4 dB in Fig. 6b when feedback from LDPC is added for each of the j users, giving

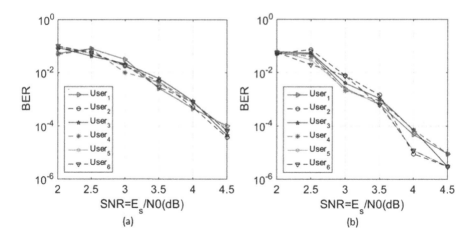

Fig. 6. The BER performance over Rayleigh fading channels with Log-MPA Decoder and LDPC Decoder (a) without any Feedback (b) with Feedback from LDPC decoder to Log-MPA

a 0.5 dB gain. Hence developed feedback scheme has shown improvements in terms of BER performance.

Using Log-domain most of the multiplication and exponential calculations are converted to simple addition or maximization. Only $M^{d_f} \times K$ multiplications remains during initialization in (7). $(K \times M^{d_f} \times d_f) + (J \times M \times d_f)$ additions with $K \times M \times d_f$ times maximization are done for each iteration. Since extrinsic information from LDPC decoder feedback, is also added to initial calculation due to fact that calculations are done in log-domain, the feedback does not adds to much of complexity for Log-MPA decode.

4   Conclusion

In this paper, we have proposed Log-MPA decoder with LDPC decoder feedback that gives better BER performance in accordance to conventional Log-MPA and LDPC Decoders. Initially we studied iterative SCMA decoding based on log-MPA detection. We worked on adaptability of output from Log-MPA so that it can be forwarded to LDPC decode. So we investigated conversion of symbol level LLR conversion from output from Log-MPA to bit level LLR. Similarly, we worked on adaptability of feedback from LDPC to Log-MPA so we converted bit level LLR to symbol level LLR. We have been able to transmit feedback from LDPC to Log-MPA decoder. Numerical results show that proposed model achieve performance gain in terms of BER over traditional SCMA receiver with LDPC Decoder.

References

1. Nikopour, H., Baligh, H.: Sparse code multiple access. In: IEEE 24th Annual International Symposium on Personal, Indoor, and Mobile Radio Communications (PIMRC). IEEE **2013**, pp. 332–336 (2013)
2. Alam, M., Zhang, Q.: Performance study of SCMA codebook design. In: 2017 IEEE Wireless Communications and Networking Conference (WCNC), pp. 1–5, March 2017
3. Yang, L., Liu, Y., Siu, Y.: Low complexity message passing algorithm for SCMA system. IEEE Commun. Lett. **20**(12), 2466–2469 (2016)
4. Wei, F., Chen, W.: Low complexity iterative receiver design for sparse code multiple access. IEEE Trans. Commun. **65**(2), 621–634 (2017)
5. Zhang, S., Xu, X., Lu, L., Wu, Y., He, G., Chen, Y.: Sparse code multiple access: an energy efficient uplink approach for 5G wireless systems. In: 2014 IEEE Global Communications Conference, pp. 4782–4787, December 2014
6. Ghaffari, A., Léonardon, M., Cassagne, A., Leroux, C., Savaria, Y.: Toward high-performance implementation of 5G SCMA algorithms. In: IEEE Access, vol. 7, pp. 10-402–10-414 (2019)
7. Yu, L., Lei, X., Fan, P., Chen, D.: An optimized design of SCMA codebook based on star-qam signaling constellations. In: 2015 International Conference on Wireless Communications Signal Processing (WCSP), pp. 1–5, October 2015
8. 3GPP: TS 38.212, Multiplexing and Channel Coding (Release 15), September 2017
9. Xiao, B., Xiao, K., Zhang, S., Chen, Z., Xia, B., Liu, H.: Iterative detection and decoding for SCMA systems with LDPC codes. In: 2015 International Conference on Wireless Communications Signal Processing (WCSP), pp. 1–5, October 2015
10. Nikopour, H., Baligh, M.: Systems and methods for sparse code multiple access, US Patent 9,240,853, 19 January 2016
11. Gallager, R.: Low-density parity-check codes. IRE Trans. Inf. Theory **8**(1), 21–28 (1962)
12. MacKay, D.J.C., Neal, R.M.: Good codes based on very sparse matrices. In: Boyd, C. (ed.) Cryptography and Coding 1995. LNCS, vol. 1025, pp. 100–111. Springer, Heidelberg (1995). https://doi.org/10.1007/3-540-60693-9_13
13. Robertson, P., Villebrun, E., Hoeher, P.: A comparison of optimal and sub-optimal map decoding algorithms operating in the log domain. In: Proceedings IEEE International Conference on Communications ICC 1995, vol. 2, pp. 1009–1013, June 1995
14. Hochwald, B.M., ten Brink, S.: Achieving near-capacity on a multiple-antenna channel. IEEE Trans. Commun. **51**(3), 389–399 (2003)
15. MacKay, D.J.C., Neal, R.M.: Near shannon limit performance of low density parity check codes. Electron. Lett. **33**(6), 457–458 (1997)
16. Zheng, H., May, W., Choi, Y.-S., Zhang, S.: Link performance abstraction for ml receivers based on RBIR metrics. US Patent 8,347,152, 1 January 2013

A Fast TDMA Schedule Based on Greedy Approach

Shuai Xiaoying[✉]

Taizhou University, Taizhou, China
xyshuai@163.com

Abstract. TDMA schedule is a complex problem with ad hoc. In order to minimize frame length, maximize channel utilization, reduce running time and increase fairness, a TDMA (M-TDMA) based on greedy selection and matrix OR is proposed. M-TDMA minimizes the frame length and maximizes the network throughput through greedy selection. It reduces the computing time by matrix OR, and increases fairness by random selection. The simulations show that the computing time and average delay of M-TDMA are significantly lower than that of TDMA schemes based on SVC. M-TDMA is also superior to TP-TDMA and N-TDMA in fairness.

Keywords: TDMA · Greedy approach · SVC · Fairness

1 Introduction

Because of using shared channel, the 1-hop and 2-hop neighbors in wireless ad hoc can't transmit simultaneously, otherwise collision will emerge. TDMA schedule is widely adopted in wireless ad hoc. Slots schedule problem is a complex problem with ad hoc. Both minimize slots of a frame and maximize the utilization in a wireless ad hoc is a NP-complete problem [1, 2]. Many TDMA schedules based on SVC (Sequential Vertex Coloring) have been proposed [3–5, 8–10]. To solve vertex coloring problem [6], Larisa Komosko used bit matrix to reduce running time [7]. To maximize channel throughput, Zhang Xizheng [10] and Haixiang Shi [11] used neural network in TDMA for ad hoc.

Generally, a wireless ad hoc has dynamic topology and other characteristics. The node is a resource constrained device. So, the computing time of TDMA algorithm for a wireless ad hoc is one of key factors. As stated above, the time complexity of existing TDMA algorithms is O (N^3). Therefore, a TDMA schedule (M-TDMA) based on greedy approach with short computing time for large scale wireless ad hoc is proposed. This paper only considers how to quickly compute slots schedule when the network topology is known. TDMA schedule can be modelled as vertex color problem. Paper [12] proposed the TDMA schedule based on greedy select and slot matrix OR. The M-TDMA minimizes the frame length and maximizes the channel utilization through greedy selection. It reduces the computing time by matrix OR, and increases fairness by random selection. The average delay of scheduling obtained by M-TDMA is lower than that of other traditional SVC schemes. M-TDMA is also superior to TP-TDMA [3] and N-TDMA [10] in fairness.

O. Habachi et al. (Eds.): UNet 2019, LNCS 12293, pp. 32–40, 2020.
https://doi.org/10.1007/978-3-030-58008-7_3

2 Model

Apply a graph G to build an ad hoc network model, as in Fig. 1. Let $G = (V, E)$, where V represents the set of nodes, $V = \{1,2,...,15\}$. $E = \{(1,2),(1,3),...,(14,15)\}$, is set of edges. The adjacency matrix A_{nn} and the 2-hop neighbor adjacency matrix B_{nn} in Fig. 1 as shown in Fig. 2.

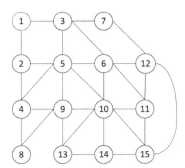

Fig. 1. 15-node topology ad hoc

```
0 1 1 0 0 0 0 0 0 0 0 0 0 0 0        0 0 0 1 1 1 1 0 0 0 0 0 0 0 0
1 0 0 1 1 0 0 0 0 0 0 0 0 0 0        0 0 1 1 1 1 0 1 1 1 0 0 0 0 0
1 0 0 0 1 1 1 0 0 0 0 0 0 0 0        0 1 0 1 1 1 0 0 1 1 1 1 0 0 0
0 1 0 0 1 0 0 1 1 0 0 0 0 0 0        1 1 1 0 1 1 0 1 1 1 0 0 1 0 0
0 1 1 1 0 1 0 0 1 1 0 0 0 0 0        1 1 1 1 0 1 1 1 1 1 1 1 1 1 1
0 0 1 0 1 0 0 0 0 1 1 1 0 0 0        1 1 1 1 1 0 1 0 1 0 1 1 1 1 1
0 0 1 0 0 0 0 0 0 0 0 1 0 0 0        1 0 0 0 1 1 0 0 0 1 0 0 0 0 1
0 0 0 1 0 0 0 0 1 0 0 0 0 0 0        0 1 0 1 1 0 0 0 1 1 0 0 1 0 0
0 0 0 1 1 0 0 1 0 1 0 0 1 0 0        0 1 0 1 1 0 0 0 1 1 0 0 1 0 0
0 0 0 0 1 1 0 0 1 0 1 0 0 1 1        0 1 1 1 1 1 0 1 0 1 1 0 1 1 1
0 0 0 0 0 1 0 0 0 1 0 1 0 0 1        0 1 1 1 1 1 0 1 1 0 1 1 0 1 1
0 0 0 0 0 1 1 0 0 0 1 0 0 0 1        0 0 1 0 1 1 1 0 1 1 0 1 1 1 1
0 0 0 0 0 0 0 0 1 1 0 0 0 1 0        0 0 1 0 1 1 0 0 0 1 1 0 0 1 1
0 0 0 0 0 0 0 0 0 1 0 0 1 0 1        0 0 0 1 1 1 0 1 1 1 1 0 0 1 1
0 0 0 0 0 0 0 0 0 1 1 1 0 1 0        0 0 0 0 1 1 0 0 1 1 1 1 1 0 1
                                     0 0 0 0 1 1 1 0 1 1 1 1 1 1 0
```

Adjacency matrix   2-hop adjacency matrix

Fig. 2. Matrix A_{nn} and B_{nn}

The elements of matrix A and matrix B are shown in (1) and (2) respectively.

$$a_{ij} = \begin{cases} 1 & \text{if node i and j are 1 - hop neighbor} \\ 0 & \text{otherwise} \end{cases} \tag{1}$$

$$b_{ij} = \begin{cases} 1 & \text{if node i and j are 2 - hop neighbor} \\ 0 & \text{otherwise} \end{cases} \tag{2}$$

The compatibility matrix D_{nn} is obtained by A OR B, $D_i = A_i \vee B_i$ for all i. The element d_{ij} as follow:

$$d_{ij} = \begin{cases} 1 & \text{if node i and j are 1 - hop or 2 - hop neighbor} \\ 0 & \text{otherwise} \end{cases} \tag{3}$$

According to Fig. 2, the 1-hop and 2-hop neighbors of node 1 are represented by A_1 and B_1 respectively, then $D_1 = A_1 \vee B_1$, as Fig. 3.

$$
\begin{array}{ll}
A_1 & 011000000000000 \\
B_1 \vee & 000111100000000 \\
\hline
D_1 & 011111100000000
\end{array}
$$

Fig. 3. $D_1 = A_1 \vee B_1$

T is n * n slot scheduling matrix representing the slot i whether can be set to node j.

$$
t_{ij} = \begin{cases} 0 & \text{if slot i can be set to node j} \\ 1 & \text{otherwise} \end{cases} \tag{4}
$$

T is initialized to 0. T_i is updated by T_i OR D_j if $t_{ij} = 0$. For example, $T_1 = 000000000000000$, $D_1 = 011111100000000$. The $t_{11} = 0$, then T_1 is updated as follow Fig. 4.

$$
\begin{array}{ll}
D_1 & 011111100000000 \\
T_1 \vee & 000000000000000 \\
\hline
T_1 & 011111100000000
\end{array}
$$

Fig. 4. T_1 update

Let S represents the slots of per frame. Let $s : S \times N \mapsto \{0, 1\}$ to express a TDMA schedule, as follow:

$$
s_{ij} = \begin{cases} 1 & \text{if slot i given to node j} \\ 0 & \text{otherwise} \end{cases} \tag{5}
$$

So, Slots schedule in wireless ad hoc can be modelled as multi-objective optimization problem.

Minimize S

$$
\text{Maximize} \sum_{i=1}^{S} \sum_{j=1}^{N} s_{ij} \tag{6}
$$

$$
s.t. \sum_{i=1}^{S} s_{ij} \geq 1 \qquad i = 1, 2, \ldots, S \tag{7}
$$

$$
\sum_{i=1}^{S} \sum_{j=1}^{N} \sum_{k=1, k \neq j}^{N} s_{ij} d_{jk} s_{ik} = 0 \tag{8}
$$

Equation (7) ensures that each node in a frame must set at least one slot. Equation (8) prevents nodes within 1-hop and 2-hop from transmitting at the same time.

3 Algorithm

Let DD_i represent the number of 1 of Di, for example $DD_1 = 6$ as Fig. 1. The element t_{ij} of matrix T indicates whether slot i can be set to node j. If $t_{ij} = 1$, node j can't use slot i. The algorithm based greedy approach consists of minimizing the frame length phase and maximizing the channel utilization phase.

3.1 Minimizing Frame Length

Each node in ad hoc is allocated to at least one slot in a frame and make the frame length as small as possible. Firstly, the nodes are listed in descending order by DD_i. If the DD_i of nodes are equal, the positions are randomly selected. The location of node i is defined by (9).

$$Loc(i) = \begin{cases} i & \text{if } DD_i \neq DD_{i+1} \\ random(i, \cdots, j) & \text{if } DD_i = DD_{i+1} = \cdots = DD_j \end{cases} \qquad (9)$$

Secondly, slot matrix T is used for slot scheduling. Starting from row 1, searching T to find an available slot in turn until all nodes are allocated. When available time slot i is found ($t_{ij} = 0$), update T_i with T_i OR D_j. The algorithm is as follows:

Input A and B;

Initialization $T = 0$;

$D = A \vee B$;

Sort nodes;

Locate nodes by (9);

Allocate slots for each node and update T by $T_i = T_i \vee D_j$;

Return length of frame

3.2 Maximum Channel Utilization

TDMA schedule can is modelled as (6). To maximum throughput, a slot is allocated to as many node as possible. Each node in ad hoc searches all the non-first available slots. For each non- first available slot, find out all nodes (candidate nodes) that the slot can be allocated. For each candidate node k, D_k OR T_i, to get a schedule that can allocate slot i to as many nodes as possible. The algorithm is as follows:

For (j=1; $j \leqslant n$; j++)   //each node

For(i=1; $i \leqslant m$;i++) //all slots

find non-first available slot i for node j

$k = agr \min_{k \in [j,n]} (T_i \vee D_k)$

$T_i = T_i \vee D_k$

$S = \neg T$ //inverse

4 Performance Indices

To evaluate performance of difference algorithms, many evaluation indices (such as running time, average delay time and fairness) is used.

4.1 Computing Time

Running time is an important metric for a TDMA scheduling algorithm. Many proposed approaches suffered from the trade-off between computing time and the optimal solution [13]. Computing time is computed on a simulator that used C program on Windows 10.

4.2 Delay

The delay time is a key factor in ad hoc. It is very important for optimizing ad hoc design. Use N/M/D/1 queues to model TDMA ad hoc, where M is the length of TDMA frame, N represents the number of nodes. Let D_{av} denote the average delay time for each node in ad hoc to broadcast packets, according to Nobuo Funabiki and Junji Kitamichi [14], D_{av} is given by:

$$D_{av} = \frac{1}{N} \times \sum_{i=1}^{N} \frac{M}{\sum\limits_{j=1}^{M} S_{ij}} = \frac{M}{N} \times \sum_{i=1}^{N} \frac{1}{\sum\limits_{j=1}^{M} S_{ij}} \tag{10}$$

Where S_{ij} denotes whether the jth slot in a frame is allocated to node i. The number of slots allocate to node i is $\sum\limits_{j=1}^{M} S_{ij}$.

Another delay time is defined by [15]. Let D_i denote the average delay of node i, as (11).

$$D_i = \frac{1}{u_i} + \frac{1}{\lambda_i} \times (\frac{\lambda_i}{u_i})^2 \times \frac{1}{2 \times (1 - \frac{\lambda_i}{u_i})} \tag{11}$$

Where λ_i is the packet arrival rate for node i, u_i (packet/slot) is the service rate of node i.

$$u_i = \sum_{m=1}^{M} \frac{S_{mi}}{M} \tag{12}$$

The total delay as follow:

$$D = \frac{\sum\limits_{i=1}^{N} \lambda_i \times D_i}{\sum\limits_{i=1}^{N} \lambda_i} \tag{13}$$

This paper adopted first delay time.

4.3 Channel Utilization

Channel utilization η is defined in [9], as (14).

$$\eta = \frac{1}{N} \sum_{j=1}^{N} \eta_j = \frac{1}{N*M} \sum_{j=1}^{N} \sum_{i=1}^{M} S_{ij} \tag{14}$$

Where η_j denotes the channel utilization for node j.

$$\eta_j = \frac{\sum_{i=1}^{M} S_{ij}}{M} \tag{15}$$

4.4 Fairness

The fairness is very important for wireless ad hoc scheduling. Unfair TDMA schedule may result in starvation of some nodes, waste of channel and others, thus reducing the performance of ad hoc. Therefore, many fair TDMA scheduling algorithms are proposed in ad hoc [15, 16]. The proposed TDMA scheduling hopes to achieve both fairness and efficiency. This paper adopts the Jain's Fairness Index [17] to measure fairness of TDMA scheduling, that is:

$$f(x) = \frac{(\sum_{i=1}^{N} s_i)^2}{N \sum_{i=1}^{N} s_i^2} \tag{16}$$

Where S_i denotes the number of slots set to node i.

5 Simulation

Assume that the global topology is known. Simulator that used C code to run on ThinkPad (Windows 10, Intel Core™ 2.6 GHz, 8 GB RAM). The running time, average delay and fairness of the M-TDMA, TP-TDMA and N-TDMA are compared in different networks such as 15-node [9], 40-node [9] and 60-node network. The 60-node network is randomly generated according to probability p. First, set $p = 0.1$ to generate a spare 60-node random ad hoc. We compared the running time and others of M-TDMA with TP-TDMA and N-TDMA. Then, the average delay and fairness of M-TDMA, TP-TDMA and N-TDMA in different 60-node random ad hoc networks are compared by simulation.

Figure 5 compares the running time of M-TDMA with TP-TDMA and N-TDMA in three networks. The simulation results show that the average running time of the TP-TDMA and N-TDMA is higher that of the M-TDMA.

Simulations show the average delay time as in Fig. 6 and Fig. 7. The p ranges from 0.05 to 0.25 to generate different random graphs of 60 nodes. The average delay of the three algorithms is shown in Fig. 7.

The fairness of the three algorithms in 15-node networks is equal. But in the other topologies, the fairness of M-TDMA is superior to TP-TDMA and N-TDMA, as shown in Fig. 8 and Fig. 9.

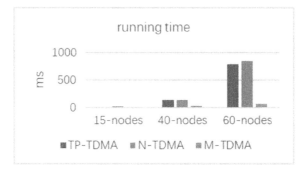

Fig. 5. Computing time of three TDMA

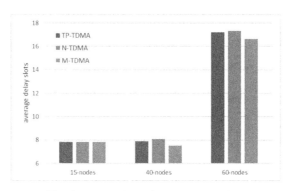

Fig. 6. Comparison of average delay time

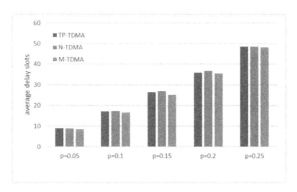

Fig. 7. Comparison of average delay time for different p

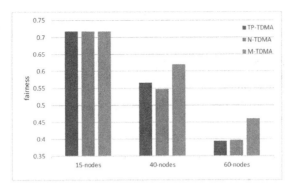

Fig. 8. Comparison of fairness

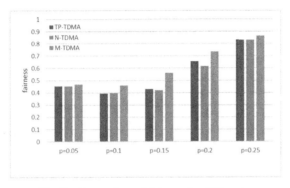

Fig. 9. Comparison of fairness for different p

6 Conclusion

In order to minimize frame length, maximize throughput, reduce running time and increase fairness, M-TDMA based on greedy approach and matrix operation is proposed. The proposed TDMA schedule uses greedy approach and matrix operation to allocate time slots. It reduces the computing time by matrix OR, and increases fairness by random selection. The time complexity of M-TDMA may reach $O(n^2)$. First, the simulations are carried out in 15-node, 40-node and 60-node random network ($p = 0.1$). Then, the average delay and fairness are simulated in different 60-node random networks generated by different p. The results show that M-TDMA is also superior to TP-TDMA and N-TDMA in average delay time and fairness.

References

1. Ephremides, A., Truong, T.V.: Scheduling broadcast in multi-hop radio networks. IEEE Trans. Commun. **38**(4), 456–460 (1990)
2. Aggeliki, S., Dimitrios, J.V., Dimitrios, D.V.: A survey of TDMA scheduling schemes in wireless multihop networks. ACM Comput. Surv. **47**(3), 1–39 (2015)

3. Jaehyun Yeo, F., Heesoo Lee, S., Sehun Kim, T.: An efficient broadcast scheduling algorithm for TDMA ad-hoc networks. Comput. Oper. Res. **29**(13), 1793–1806 (2002)
4. Clayton, W., Commander, F., Panos, M., Pardalos, S.: A combinatorial algorithm for the TDMA message scheduling problem. Computat. Optimizat. Appl. **43**(3), 449–460 (2009)
5. Sinem, C.E., Pravin, V.: TDMA scheduling algorithms for wireless sensor networks. Wireless NetWork **16**, 985–997 (2010)
6. Enrico Malaguti, F., Paolo Toth, S.: A survey on vertex coloring problems. Int. Trans. Oper. Res. **17**(1), 1–34 (2009)
7. Komosko, L., Batsyn, M., Segundo, P.S., Pardalos, P.M.: A fast greedy sequential heuristic for the vertex colouring problem based on bitwise operations. J. Combinat. Optim. **31**(4), 1665–1677 (2015). https://doi.org/10.1007/s10878-015-9862-1
8. Wang, G., Ansari, N.: Optimal broadcast scheduling in packet radio networks using mean field annealing. IEEE J. Select. Areas Commun. **15**(2), 250–260(1997)
9. Wang Lingzhi, F., Chen Lingyu, S., Yuan Aisha, T.: N-time division multiple access dynamic slot allocation protocol based on I-sequential vertex coloring algorithm. Comput. Eng. **42**(1), 89–94 (2016)
10. Zhang Xizheng, F., Wang Yaonan, S.: New mixed broadcast scheduling approach using neural networks and graph coloring in wireless sensor network. J. Syst. Eng. Electron. **20**(1), 185–191 (2009)
11. Haixiang Shi, F., Lipo Wang, S.: Broadcast scheduling in wireless multihop networks using a neural network based hybrid algorithm. Neural Netw. **18**(5), 765–771 (2005)
12. Shuai, X.: A greedy approach for TDMA based on matrix operation. In: Proceedings of CyberC, pp. 284–287 (2019)
13. Arivudainambi, D., Rekha, D.: An evolutionary algorithm for broadcast scheduling in wireless multihop networks. Wireless Netw. **18**, 787–798 (2012)
14. Nobuo, F., Junji, K.: A gradual neural network algorithm for broadcast scheduling problems in packet radio networks. IEICE Trans Fundam. **E82-A**(5), 815–824(1999)
15. Jun, H., Hung, K.P.: Fairness of medium access control protocols for multi-hop ad hoc wireless networks. Comput. Netw. **48**, 867–890 (2005)
16. Shi, H., Venkatesha Prasad, R., Onur, E. et al.: Fairness in wireless networks issues, measures and challenges. IEEE Commun. Surv. Tutorials **16**(1), 5–24(2014)
17. Raj, J.: The Art of Computer Systems Performance Analysis: Techniques for Experimental Design, Measurement, Simulation, and Modeling. Wiley, Hoboken (1991)

Extended Low Rank Parity Check Codes and Their Efficient Decoding for Multisource Wireless Sensor Networks

Nicolas Aragon[(⊠)], Jean Pierre Cances[(⊠)], Imad El Qachchach[(⊠)], Philippe Gaborit[(⊠)], and Oussama Habachi[(⊠)]

Xlim Institute of Technology, University of Limoges, Limoges, France
{nicolas.aragon,jean-pierre.cances,el-qachchach,
philippe.gaborit,oussama.habachi}@xlim.fr

Abstract. In this paper, we consider a multisource network transmitting information through relays to a base station using Network Coding. We design a model for this scenario and use the rank metric to address the problem of packet errors (caused for example by a malicious user or a defective node). We introduce a new family of codes, the extended LRPC codes, that are very well suited to this model and extensively use the fact that the information comes from multiple sources to decode. They therefore improve the communication reliability compared to classical LRPC codes and Gabidulin codes. We provide a theoretical analysis of their decoding failure probability, both in a one source and multisource scenario, as well as simulation results confirming our analysis.

1 Introduction

Network coding (NC) has been recently introduced to reduce the traffic in general networks. Plenty of works have investigated this idea in both wired and wireless networks. Indeed, NC is proved to be an appropriate solution increasing data throughput and reducing energy consumption for WSNs. NC was first introduced in the seminal paper [1] and since, it has been shown to significantly improve network efficiency by reducing the number of transmissions. Random linear network coding (RLNC) [8] is a class of network coding that uses a linear code generated randomly by every node of the network. It assumes that the data are vectors over a finite field and that each node of the network performs a random linear combination of all the received packets so far and forwards them to nearby nodes. Nevertheless, if packet error occurs, the erroneous packets are combined with unharmed ones causing the whole combination to be affected. This kind of errors can be illustrated in three use-cases. The first use-case is when a malicious user injects erroneous packets into the network to disrupt the overall system, such as the scenarios studied in [3] and [7]. The second use-case is depicted by the presence of a node failure within the network, see [6]. The third case is when we take into consideration the impact of background noise that is

© Springer Nature Switzerland AG 2020
O. Habachi et al. (Eds.): UNet 2019, LNCS 12293, pp. 41–55, 2020.
https://doi.org/10.1007/978-3-030-58008-7_4

caused by propagation channel and electronic impairment (additive white Gaussian noise (AWGN) for example). In order to solve the problem of background noise, we propose to use convolutional codes. Each node uses a linear combination of the received packets and decodes them using convolutional decoder. The first and the second cases can be solved by using rank metric codes. It has been proven that rank codes are efficient against rank errors [10]. In particular, Gabidulin proposed a class of correcting codes named *Gabidulin codes* in order to apply them for correcting criss-cross errors. A class of rank metric codes has been proposed in [5], called Low Rank Parity Check (LRPC), that has approximately the same performance of Gabidulin codes. Koetter and Kschischang tested the performance of rank codes combined with RLNC schemes for intentional attacks [9].

In this paper, we investigate existing solution in [2] for multisource networks using error correcting codes and we propose a generalized solution. We also introduce a family of codes, the LRPC codes, that are very well suited to this model.

The main contribution of this paper is a new decoding algorithm based on LRPC codes, that features a probabilistic decoding algorithm whose decoding failure rate gets really low when using multiple sources. We also derive theoretical expression of failure decoding probability at the destination for LRPC in multisource networks. Finally, we validate the theoretical results with simulations and we show that our proposition achieves good performance compared to existing ones. The simulations illustrate the advantages of using modified LRPC codes compared to classical LRPC and Gabidulin codes.

The remainder of this paper is organized as follows. In Sect. 2, notations and fundamental preliminaries of finite field and vector spaces are detailed. A detailed description of rank codes is provided in Sect. 3. Section 4 describes the system model and formulates the problem statement. The framework of the calculation of the failure decoding probability and the description of modified LRPC are expressed in Sect. 5. In Sect. 6, we present the simulation results and the conclusions are drawn in Sect. 7.

2   Preliminaries

Let q be a power of prime number p and $\mathbf{u}$ be an element of $\mathbb{F}_{q^m} \setminus \mathbb{F}_q$. In this paper, all coefficients of a vector are in the finite field $\mathbb{F}_{q^m}$. Let $\mathbb{F}_q^{m \times N}$ denote the set of all $m \times N$ matrices over $\mathbb{F}_q$ such that $m \geq N$ and let $\mathbf{b} = \{b_1, b_2, \ldots, b_m\}$ be a basis of $\mathbb{F}_{q^m}$ over $\mathbb{F}_q$.

Let $(x_1, \ldots x_n)$ n elements of $\mathbb{F}_{q^m}$. The $\mathbb{F}_q$-subspace generated by these elements is denoted $\langle x_1, \ldots, x_n \rangle$. If E and F are two subspaces of $\mathbb{F}_{q^m}^N$, then $\langle E.F \rangle$ denotes the subspace generated by the product of elements of E and F, ie $\langle E.F \rangle = \langle e_i f_j \rangle$ where the (e_i) (respectively the (f_j)) are a basis of E (respectively F). If X is a matrix in $\mathbb{F}_q^{m \times N}$, the row space of a matrix X is denoted by $\langle X \rangle$.

As it has been shown in [4], the number of t-dimensional subspace of an m-dimensional vector space over $\mathbb{F}_q$ is the Gaussian coefficient calculated by

$$\begin{bmatrix} m \\ t \end{bmatrix} \triangleq \prod_{i=0}^{t-1} \frac{q^m - q^i}{q^t - q^i}. \tag{1}$$

Hence, we can deduce from Eq. (1) the number of matrices of rank t in the space $\mathbb{F}_q^{m \times N}$, which is

$$S(m, N, q, t) = \prod_{i=0}^{t-1} \frac{(q^m - q^i)(q^N - q^i)}{q^t - q^i}. \tag{2}$$

Let Y_1 and Y_2 be two $m \times N$ matrices over $\mathbb{F}_q$. The row space of a matrix Y_1 is denoted by $\langle Y_1 \rangle$. It means that the the space $\langle Y_1 \rangle$ is generated by the rows of the matrix Y_1. Then, we have

$$\left\langle \begin{bmatrix} Y_1 \\ Y_2 \end{bmatrix} \right\rangle = \langle Y_1 \rangle + \langle Y_2 \rangle. \tag{3}$$

Therefore

$$\begin{aligned} rank \begin{bmatrix} Y_1 \\ Y_2 \end{bmatrix} &= dim\left(\langle Y_1 \rangle + \langle Y_2 \rangle \right) \\ &= rank(Y_1) + rank(Y_2) - dim(\langle Y_1 \rangle \cap \langle Y_2 \rangle). \end{aligned} \tag{4}$$

Let $\mathbf{u}$ be an element of $\mathbb{F}_{q^m} \setminus \mathbb{F}_q$ and E be a subspace of $\mathbb{F}_{q^m}$ of dimension r over $\mathbb{F}_q$. We suppose that $2r \ll m$ and we investigate the typical dimension of the subspace $E + \mathbf{u}E$. We rely on the following observation:

Proposition 1. *The probability that $E + \mathbf{u}E$ is of dimension $2r$ is given by*

$$\mathbb{P}(dim(E + \mathbf{u}E) = 2r) \approx 1 - \frac{q^{2r} - q^{r+1}}{q^m - q}.$$

Proof. Let us take a fixed r-dimensional subspace E in $\mathbb{F}_{q^m}$. Suppose that the dimension of $E + \mathbf{u}E$ is less than $2r$ for $\mathbf{u}$ randomly chosen in $\mathbb{F}_{q^m} \setminus \mathbb{F}_q$. It means that: $\exists (e_1, e_2) \in E^2$, that verifies $\mathbf{u}e_2 = e_1$. Now, we compute the number of possibilities of choosing $\mathbf{u}$ that verifies $\mathbf{u} = e_1 e_2^{-1}$, for $(e_1, e_2) \in E^2$. The number of possible values of (e_1, e_2) is at most q^{2r} and since $\mathbf{u}$ is not in $\mathbb{F}_q$ the case $(\alpha e, e)$ for $\alpha \in \mathbb{F}_q$ and $e \in E$ is not possible. Thus, the number of possibilities to choose $\mathbf{u}$ that verifies $\mathbf{u} = e_1 e_2^{-1}$, for $(e_1, e_2) \in E^2$ is $q^{2r} - q^{r+1}$. The number of possible values of $\mathbf{u}$ is $q^m - q$. $\square$

Let A be a matrix in $\mathbb{F}_q^{2r \times N-k}$ and suppose that $2r \le N - k$. By using (2), the probability that A is a full rank matrix is given by

$$\mathbb{P}(rank(A) = 2r) = \prod_{i=0}^{2r-1} (1 - q^{i-(N-k)}). \tag{5}$$

Let E be a subspace of dimension r over $\mathbb{F}_q$. Let s be a vector in $E + \mathbf{u}E$ of length $N - k$. We have the following proposition:

Proposition 2. *The probability that the subspace $\langle s \rangle$ is of dimension $2r$ over $\mathbb{F}_q$ is given by*

$$\mathbb{P}(dim(\langle s \rangle) = 2r) \approx \left(1 - \frac{q^{2r} - q^{r+1}}{q^m - q}\right) \prod_{i=0}^{2r-1} (1 - q^{i-(N-k)}).$$

Proof. Suppose that dimension of $E + \mathbf{u}E$ is $2r$ and let $\{E_1, E_2, ..., E_r, \mathbf{u}E_1, \mathbf{u}E_2, ..., \mathbf{u}E_r\}$ be a basis of $E + \mathbf{u}E$. All coefficients of the vector s are in $E + \mathbf{u}E$ by definition of s. The vector s can be written as follows:

$$s = (E_1, ..., E_r, \mathbf{u}E_1, ..., \mathbf{u}E_r) \times A,$$

where, A is a matrix in $\mathbb{F}_q^{2r \times N-k}$. Since the coefficients of s are random elements of $E + \mathbf{u}E$, the matrix A is also random. The probability that the set of all coefficients of s generates the whole space is the probability that A is a full rank matrix. From Eq. (5), the probability that a random matrix A is full rank is $\prod_{i=0}^{2r-1} (1 - q^{i-(N-k)})$. Now, the probability that $dim(E + \mathbf{u}E) = 2r$ is given in the Proposition 1. $\qquad\square$

It is interesting to remark that in practice the probability $\mathbb{P}(dim(E + \mathbf{u}E) = dr)$ decreases much more faster to 0 when $dr \ll m$. Thus, the probability that $dim(\langle s \rangle) = dr$ given in the previous proposition can be approximated by:

$$\mathbb{P}(dim(\langle s \rangle) = dr) \approx \prod_{i=0}^{dr-1} (1 - q^{i-(N-k)}). \tag{6}$$

3   Rank Metric

In this section, we present some concepts from rank metric coding theory. The reader is referred to [5] and [4] and references therein for further details. A brief overview of concepts relevant to this work can be found in [12]. Afterwards, we introduce Gabidulin codes and LRPC codes, and then we propose modified decoding algorithm of LRPC.

Let $\mathbf{v}$ be a vector of $\mathbb{F}_{q^m}^N$. For $i \in \{1, 2, ..., N\}$, we have $v_i = \sum_{j=1}^{m} v_{ij} b_j$ and $\mathbf{v}$ can be interpreted as a matrix $V = (v_{ij}) \in \mathbb{F}_q^{m \times N}$. We can define the rank weight of $\mathbf{v}$ over $\mathbb{F}_q$ as the rank of the associated matrix V, denoted $rank(\mathbf{v})$. The rank distance between two vectors v and w of $\mathbb{F}_{q^m}^N$ is defined by $d_r(\mathbf{v}, \mathbf{w}) = rank(\mathbf{v} - \mathbf{w})$. These definitions are independent of the choice of the basis $\{b_1, b_2, ..., b_m\}$.

We can now define the support of a vector. This definition differs from the Hamming metric:

Definition 3. *Let $v \in \mathbb{F}_{q^m}^N$. The support of v is the $\mathbb{F}_q$ − subspace of $\mathbb{F}_{q^m}$ generated by its coordinates:*

$$Supp(v) = \langle v_1, ..., v_N \rangle$$

Definition 4. *A rank code C of length N and dimension k over $\mathbb{F}_{q^m}$ is a subspace of dimension k of $\mathbb{F}_{q^m}^N$ equipped with the rank metric.*

Similar to the minimum Hamming distance for linear codes we define the minimum rank distance of a code C.

Definition 5. *The minimum rank distance of a code C is given by:*

$$d_r^m = min\{rank(\boldsymbol{v}) \mid \boldsymbol{v} \in C, \boldsymbol{v} \neq 0\}.$$

3.1 Gabidulin Codes

Gabidulin codes are introduced in [4], the well-known class of Maximum Rank Distance (MRD) codes. They have been already used successfully in many applications such as cryptography [5], power-line communications [12] and network coding [9].

The Gabidulin code of length N, dimension k and support $g = (g_1, g_2, \ldots, g_N)$ is the set of words obtained by evaluating q-polynomials of q-degree at most $k - 1$ at g_1, g_2 and g_N.

$$Gab(g, k, N) = \{(P(g_1), \ldots, P(g_N)) \mid deg_q(P) \leq k - 1\}.$$

The decoding of Gabidulin codes can be done based on q-polynomials by using modified Berlekamp-Massey algorithm [11] or extended euclidean algorithm in the non-commutative ring of q-polynomial. They can decode errors of weight up to $\lfloor \frac{N-k}{2} \rfloor$ without probability of failure.

3.2 Low Rank Parity Check Codes

The LRPC code and its parity check matrix are described in the following definition.

Definition 6. *A Low Rank Parity Check code of low rank d, length N and dimension k and with a parity check matrix $\boldsymbol{H} = (h_{ij})$ over $\mathbb{F}_{q^m}$ such that the sub-vector space of $\mathbb{F}_{q^m}$, generated by the coefficients h_{ij} of the matrix $\boldsymbol{H}$, has dimension equals to d.*

Without loss of generality, in this article we are interested in the case $d = 2$. Let $M = (m_{ij})$ be a lower triangular matrix in $\mathbb{F}_q^{2(N-k) \times N}$ and let F be a subspace of $\mathbb{F}_{q^m}$ of dimension 2 generated by the basis $\{1, \mathbf{u}\}$. The matrix $\boldsymbol{H} = (h_{ij})$ is constructed such that $h_{ij} \in F$. Then, for $1 \leq i \leq N - k, 1 \leq j \leq N$, $h_{ij} = h_{ij1} + \mathbf{u}h_{ij2}$, where h_{ij1} and h_{ij2} are elements of $\mathbb{F}_q$. In order to reduce the complexity of decoding the LRPC codes, we set $h_{ij1} = m_{(2i-1),j}$ and $h_{ij2} = m_{2i,j}$, for $1 \leq i \leq N - k$ and $1 \leq j \leq N$.

Suppose that the error $(e_1, ..., e_N)$ is of weight r and e_i are elements of the error space E of dimension r generated by a basis $\{E_1, E_2, \cdots, E_r\}$. Then, all $e_i(1 \leq i \leq N)$ can be written as $e_i = \sum_{j=1}^{r} e_{ij}E_j$. Suppose that the dimension of the space $E + \mathbf{u}E$ is exactly $2r$ (see Proposition 1). It is then possible to express the system of equations $\mathbf{H}.e^T = s$ over $\mathbb{F}_{q^m}$ into system of equations over $\mathbb{F}_q$, by expressing the syndrome coordinates in the product basis $\{E_1, .., E_r, \mathbf{u}E_1, ..., \mathbf{u}E_r\}$, for $1 \leq i \leq N - k$, as follows:

$$s_i = \sum_{k=1}^{r} s_{i1k}E_k + \mathbf{u}\sum_{k=1}^{r} s_{i2k}E_k.$$

We have $\mathbf{A}_H^r.e'^T = s'$, where $e' = (e_{11}, ..., e_{1r}, e_{21}, ..., e_{nr})$ and $s' = (s_{111}, ..., s_{11r}, ..., s_{(n-k)2r})$. We have detailed the matrix A_H^r in a previous work (see [12]).

The decoding algorithm can fail if the support of s is of dimension strictly smaller than $2r$. Thus we have the following proposition:

Proposition 7. *An LRPC code of rank d, length N and dimension k can decode errors of weight up to $\lfloor \frac{N-k}{2} \rfloor$ with probability of $\approx 1 - q^{N-k+1-dr}$, where r is the rank of the error.*

Proof. According to Eq. (6), we have

$$\mathbb{P}(dim(\langle s \rangle) = dr - 1) \approx \prod_{i=0}^{dr-1}(1 - q^{i-(N-k)})$$

$$\approx 1 - q^{-(N-k+1-dr)}$$

$\square$

Algorithm 1: Decoding algorithm of the LRPC codes

Input: The parity check matrix $\boldsymbol{H}$, the syndrome $\boldsymbol{s}$
Output: The error vector $\boldsymbol{e}$ of rank r
1 $S \leftarrow < s_1, \ldots, s_{n-k} >$
2 $E \leftarrow F_1^{-1}.S \cap F_2^{-1}.S$, where $\{F_1, F_2\}$ is a basis of F
3 Try solving $\boldsymbol{H}.e^t = \boldsymbol{s}$ with $\boldsymbol{e} \in E^n$
4 **return** $\boldsymbol{e}$

3.3 Decoding Algorithm

In the following we only consider LRPC codes of rank 2.

3.4 Probability of Failure

In order to estimate the decoding failure rate of this algorithm, we need to study the probability that we do not recover the support E of the error. Since we can choose the parity check matrix $\boldsymbol{H}$ such that the system $\boldsymbol{H}.e^t = \boldsymbol{s}$ is invertible, this can not be a source of failure.

Proposition 8. *An $[n, k]$ LRPC code of rank 2 can decode errors of rank r up to $\lfloor \frac{k}{2} \rfloor$ with probability :*

$$(1 - \frac{S(k, 2r, q, 2r)}{q^{2r(k)}})$$

We use the expression for the failure decoding probability given in Theorem 11 and compare the resulting values with the simulation results. Figure 4 depicts simulated (S) and theoretical (T) expression of successful decoding probability for $d = 2, n = 24, k = 15, s = 1$ and $t = 6$ as a function of the number of erroneous packets. It can be observed that the system performance is close to the formula given in Theorem 11 (Fig. 1).

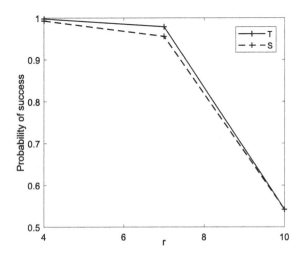

Fig. 1. Simulation results for $d = 2, n = 24$ and $k = 15$.

3.5   Multisource Case

Theorem 9. *Using syndromes from N sources, the LRPC codes can decode errors of rank r up to $r \leqslant \lfloor \frac{Nk}{2} \rfloor$ with probability :*

$$(1 - \frac{S(Nk, 2r, q, 2r)}{q^{2r(Nk)}}).$$

4   System Model and Problem Formulation

We consider a network comprising a base station BS, s source nodes $\mathcal{S}_1, \mathcal{S}_2, ..., \mathcal{S}_s$ and a number of relay nodes. Each source node is attempting to transmit m packets to the BS through relay nodes, as illustrated in Fig. 2.

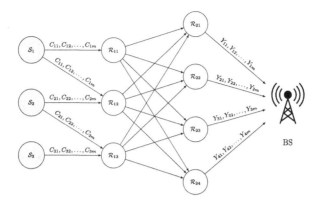

Fig. 2. Example of a network composed of 3 sources, number of relay nodes and BS.

To this end, the source S_i segments data into m packets $u_{i1}, u_{i2}, ..., u_{im}$ of length k, then encodes them using a rank code and transmits the coded packets to the relay nodes. Let $C_{i1}, C_{i2}, ..., C_{im}$ denote the coded packets of node S_i. Hence, $S_1, S_2, ..., S_s$ transmit $m \times s$ coded packets of length N to the relay nodes. Each relay node that receives the source packets employs RLNC to combine them and generates coded packets. Note that the coefficients are randomly chosen from $\mathbb{F}_q$, where q is the field size. Afterwards, relays send the generated packets to other relays until the coded packets are received by the destination BS. Let $Y_{11}, Y_{12}, ..., Y_{sm}$ denote the received packets which can be expressed in s block matrices of size $(m \times N)$.

We consider the application of Physical-layer Network Noding (PNC) between the relay nodes as shown in Fig. 3. Each stage of the network behaves as independent network and differently of other stages. In this model, relays $\mathcal{N}_1, \mathcal{N}_2, ..., \mathcal{N}_l$ send information to a node $\mathcal{N}$ in the next stage. We assume that all nodes are half-duplex. The first time slot corresponds to an uplink phase, in which nodes $\mathcal{N}_1, \mathcal{N}_2, ..., \mathcal{N}_l$ transmit their coded packets simultaneously to the node $\mathcal{N}$. The node $\mathcal{N}$ then constructs a network coded packet based on the simultaneously received signals from $\mathcal{N}_1, \mathcal{N}_2, ..., \mathcal{N}_l$. The second time slot corresponds to a downlink phase, in which $\mathcal{N}$ attempts to recover the original packet transmitted by $\mathcal{N}_1, \mathcal{N}_2, ..., \mathcal{N}_l$ and sends it to next stage nodes.

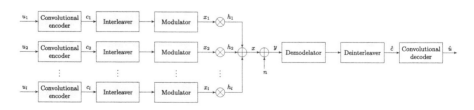

Fig. 3. The system model for the inner code.

In the following, we focus on improving the error decoding performance of convolutional code. As shown in Fig. 3, nodes $\mathcal{N}_1, \mathcal{N}_2, ..., \mathcal{N}_l$ adopt the same convolutional code with length N and k. In this paper, nodes use the same pseudo-random bit-interleaver instead of the conventional bit-interleaver to allocate the coded bits to different modulation levels. Without loss of generality, we focus on BPSK modulation. Our framework can be easily extended to higher order constellations. We assume that the power control and the synchronization at all nodes are perfect.

Consider transmission of l packets to the node $\mathcal{N}$. The received packet is:

$$y = (x_1 h_1 + n_1) + (x_2 h_2 + n_2) + \cdots + (x_l h_l + n_l), \tag{7}$$

where h_i is the channel coefficients of the channels between the node $\mathcal{N}_i$ and the node $\mathcal{N}$. It can be considered as an $N \times N$ diagonal matrix where diagonal coefficients have a Rayleigh distribution with parameter $\sigma = \sqrt{\frac{1}{2}}$. The parameter $n = n_1 + n_2 + \cdots + n_l$ represents the channel additive Gaussian noise (AWGN), where $n_1, n_2, ..., n_l$ are independent Gaussian variables with zero mean and variance $\sigma_1^2 = \sigma_2^2 = \cdots = \frac{N_0}{2}$; i.e. $n \sim \mathcal{N}(0, \frac{mN_0}{2})$.

In order to limit the impact of background noises that are caused by the nature of the wireless channel, we use a convolutional code. Each relay node verifies the integrity of the received packets. If the received packets is erroneous, the node uses convolutional decoder in order to recover the transmitted packet. However, if we combine a big number of packets the total variance of the noise increases significantly and then the convolutional decoder cannot recover the correct codeword. Also, packets generated by malicious nodes cannot be detected by the convolutional since the latter can use convolutional code too. In this case, relay node that receives the wrong packets combine them with the correct ones generating a wrong packet too. Let r denote the number of erroneous packets caused by the combination of a big number of received packets.

Suppose that r erroneous packets are injected into the network during the transmission of the $m \times s$ source packets. Since packets are randomly combined, errors may affect all the packets. Particularly, errors may affect all the packets of one source. At the BS, the packets of each source are put together in order to apply the rank decoder. By using a classical rank code, the decoding algorithm uses the information of m received packets so as to recover the source packets. For a particular source, if r is bigger than m, the rank error may be bigger than the decoding capability of the rank code. Thus, the BS cannot recover the source packets.

The main idea of this paper is to use the error information of all received packets in order to recover the error basis. Then, we use the error basis in the decoding algorithm to recover packets of each source.

5   Extended LRPC Codes

5.1   Definition and Decoding Algorithm

Definition 10. *Extended LRPC codes*
An $[n+t, k]$ extended LRPC code of rank d over $\mathbb{F}_{q^m}$ is a code such that it has a parity check matrix H consisting of an $n \times (n-k)$ parity check matrix of an LRPC code, extended by an identity matrix of size t on the first coordinates:

$$H = \begin{pmatrix} I_t & 0 \\ 0 & H_{LRPC} \end{pmatrix}.$$

The probabilistic decoding algorithm of this family of codes is an adaptation of the decoding algorithm of the LRPC codes, to use the fact that the first syndrome coordinates are actually coordinates of the error. In the following we only consider extended LRPC codes of rank 2.

Algorithm 2: Decoding algorithm of the extended LRPC codes

Input: The parity check matrix H, the syndrome s
Output: The error vector e of rank r
1 $E' \leftarrow < s_1, \ldots, s_t >$
2 $S \leftarrow < E'.F > + < s_{t+1}, \ldots, s_{n-k+t} >$
3 $E \leftarrow F_1^{-1}.S \cap F_2^{-1}.S$, where $\{F_1, F_2\}$ is a basis of F
4 Try solving $H.e^t = s$ with $e \in E^{n+t}$
5 **return** e

5.2   Probability of Failure

In order to estimate the decoding failure rate of this algorithm, we need to study the probability that we do not recover the support E of the error. Since we can choose the parity check matrix H such that the system $H.e^t = s$ is invertible, this can not be a source of failure.

Theorem 11. *An $[n+t, k]$ extended LRPC code of rank 2 can decode errors of rank r up to $\lfloor \frac{2t+k}{2} \rfloor$ with probability:*

$$\sum_{j=0}^{min(r-1,t)} \frac{S(t, r, q, j)}{q^{rt}} \times (1 - \frac{S(k+2j, 2r, q, 2r)}{q^{2r(k+2j)}})$$

Proof. The probability that the first t coordinates of the syndrome span a subspace of dimension j of E is equal to the number of matrices of size $t \times r$ of rank j over $\mathbb{F}_q$ divided by the total number of matrices of size $t \times r$ over

$\mathbb{F}_q : \frac{S(t,r,q,j)}{q^{rt}}$. If the dimension is exactly r, then the algorithm will succeed. For each other potential dimension, we need to study the probability that $<E'.F> + <s_{t+1},\ldots,s_{n-k+t}>$ span the whole product space $\langle E.F \rangle$.

If we write $<E'.F> + <s_{t+1},\ldots,s_{n-k+t}>$ as a $k + 2j \times 2r$ matrix over $\mathbb{F}_q$, then the probability that these vectors do not span the whole space $<E'.F>$ is $1 - \frac{S(k+2j,2r,q,2r)}{q^{2r(k+2j)}}$, hence the result.

□

We use the expression for the failure decoding probability given in Theorem 11 and compare the resulting values with the simulation results. Fig. 4 depicts simulated (S) and theoretical (T) expression of successful decoding probability for $d = 2, n = 24, k = 15, s = 1$ and $t = 6$ as a function of the number of erroneous packets. It can be observed that the system performance is close to the formula given in Theorem 11.

5.3   Multisource Case

Theorem 12. *Using syndromes from N sources, the extended LRPC codes can decode errors of rank r up to $r \leqslant \lfloor \frac{2Nt+Nk}{2} \rfloor$ with probability:*

$$\sum_{j=0}^{min(r-1,Nt)} \frac{S(Nt,r,q,j)}{q^{rNt}} \times (1 - \frac{S(Nk+2j,2r,q,2r)}{q^{2r(Nk+2j)}}).$$

Proof. The proof is similar to the proof of Theorem 11, except that we get Nt elements of the vector space E, and Nk elements of $\langle E.F \rangle$ in the syndrome coordinates.

□

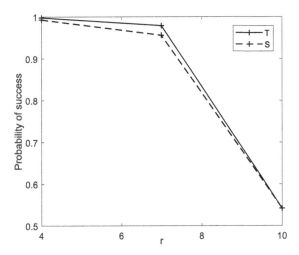

Fig. 4. Simulation results for $d = 2, n = 24, k = 15$ and $t = 6$.

6   Numerical Results

In this section, we investigate the performance of the proposed model via simulation and compare the results of the proposed modified LRPC code with the classical LRPC code. First, we test the behavior of the two codes in the absence of AWGN noise and then, we evaluate the impact of background noise on both codes.

6.1   The Comparison Between Modified LRPC and Classical LRPC in the Absence of AWGN

We set the number of source packets to 80 and the number of source nodes to 1, 2 and 3 respectively. The source coded packets have the same length n. The relevant dimensions of the parity-check matrix are $n = 17$, $k = 10$ and $d = 2$. We use a binary phase shift keying (BPSK).

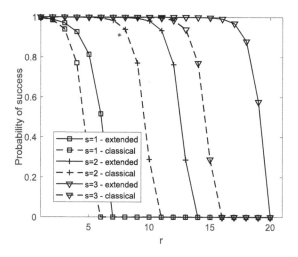

Fig. 5. Probability of successful decoding for $n = 17$, $k = 10$ using 1, 2 and 3 sources.

Figure 5 illustrates the probability of successful decoding as a function of the number of erroneous packets injected into the network for different numbers of sources. It can be observed that modified LRPC has a good behavior compared to the classical LRPC. By increasing the number of sources, the gap between the two graphs becomes increasingly important. This is because extended LRPC code has $s \times t/2$ additional information of the error support that uses in the decoding process.

6.2 The Comparison Between Extended LRPC and Classical LRPC in the Presence of AWGN

In the second experiment we compare the performance of the extended LRPC and classical LRPC in the presence of additive white Gaussian noise. We fix the number of erroneous packets injected into the network to 4.

We use extended LRPC and classical LRPC as *outer codes*. Then, the coded packets are coded again using convolutional code at the source nodes, and transmitted to the next relays. At the intermediate levels, we use the classical RLNC. For convolutional encoder, with a standard $rate = \frac{1}{2}$ and $K = 7$, we use an interleaver to improve the error correction.

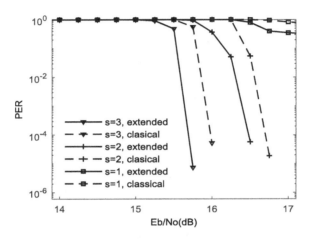

Fig. 6. Packet Error Rate as a function of SNR for $r = 4$ using 1, 2 and 3 sources.

We can observe, in Fig. 6, that the extended LRPC is about 0.4 dB better than the classical LRPC. The use of a rank code does not have a beneficial contribution regards to the channel errors. This is because of the property of the white noise, each symbol has a big probability to generate a rank error and therefore reducing the error-correction capability. This is the reason of using a convolutional code to reduce the channel errors impact. It is obvious that the performance of both rank codes deteriorate for $s = 1$ this is because the decoding failure probability of extended LRPC and classical LRPC are affected by the rank error in the case when $s = 1$.

7 Conclusion

In this paper, we proposed a new family of LRPC codes, extended LRPC codes, which are particularly well suited for use in multisource network using RLNC. We propose a new decoding algorithm that takes into account the fact that the information comes from multiples sources, which is not possible when using Gabidulin codes, and reduces the decoding failure rate over the classical LRPC codes.

The considered scenario takes into account not only errors caused by the nature of the wireless channel, but also errors introduced by a malicious users or due to node failures. In fact, we use extended LRPC as an outer code and we use the convolutional code as an inner code to deal with the wireless channel errors. We have derived analytically the exact expression for the decoding probability of extended LRPC codes. Numerical results have shown that both the simulation and the theoretical expression for the decoding probability of extended LRPC codes are very tight and accurately predict the decoding probability. Our analysis has also exposed the clear benefits of the extended LRPC in terms of recovery accuracy compared to both the classical LRPC codes and the Gabidulin codes.

References

1. Ahlswede, R., Cai, N., Li, S.Y.R., Yeung, R.W.: Network information flow. IEEE Trans. Inf. Theory **46**(4), 1204–1216 (2000)
2. El Qachchach, I., Habachi, O., Cances, J.P., Meghdadi, V.: Efficient multi-source network coding using low rank parity check code. In: 2018 IEEE Wireless Communications and Networking Conference (WCNC), pp. 1–6. IEEE (2018)
3. Fiandrotti, A., Gaeta, R., Grangetto, M.: Simple countermeasures to mitigate the effect of pollution attack in network coding-based peer-to-peer live streaming. IEEE Trans. Multimed. **17**(4), 562–573 (2015). https://doi.org/10.1109/TMM. 2015.2402516
4. Gabidulin, E.M.: Theory of codes with maximum rank distance. Probl. Inf. Transm. (English translation of Problemy Peredachi Informatsii) **21**(1), 3–16 (1985)
5. Gaborit, P., Murat, G., Ruatta, O., Zémor, G.: Low rank parity check codes and their application to cryptography. In: Proceedings of the WCC, pp. 168–180 (2013)
6. Ho, T., Koetter, R., Medard, M., Karger, D.R., Effros, M.: The benefits of coding over routing in a randomized setting. In: Proceedings of the IEEE International Symposium on Information Theory, pp. 442+. IEEE, June 2003. https://doi.org/ 10.1109/isit.2003.1228459
7. Ho, T., Leong, B., Koetter, R., Medard, M., Effros, M., Karger, D.R.: Byzantine modification detection in multicast networks with random network coding. IEEE Trans. Inf. Theory **54**(6), 2798–2803 (2008). https://doi.org/10.1109/TIT.2008. 921894
8. Ho, T., Koetter, R., Medard, M., Karger, D.R., Effros, M.: The benefits of coding over routing in a randomized setting (2003)
9. Koetter, R., Kschischang, F.R.: Coding for errors and erasures in random network coding. IEEE Trans. Inf. Theory **54**(8), 3579–3591 (2008). https://doi.org/ 10.1109/TIT.2008.926449
10. Plass, S., Richter, G., Vinck, A.H.: Coding schemes for crisscross error patterns. Wirel. Pers. Commun. **47**(1), 39–49 (2008). https://doi.org/10.1007/s11277-007-9389-6

11. Richter, G., Plass, S.: Fast decoding of rank-codes with rank errors and column erasures. In: Proceedings of the International Symposium on Information Theory, ISIT 2004, pp. 398–398. IEEE (2004)
12. Yazbek, A.K., El Qachchach, I., Cances, J.P., Meghdadi, V.: Low rank parity check codes and their application in power line communications smart grid networks. Int. J. Commun. Syst. **30**, 1–9 (2017)

Analysis of the Coexistence of Ultra Narrow Band and Spread Spectrum Technologies in ISM Bands

Mohamed Amine Ben Temim$^{(\boxtimes)}$, Guillaume Ferré, and Romain Tajan

Univ. Bordeaux, Bordeaux INP, CNRS, IMS, UMR 5218,
351 Cours de la Libération, 33400 Talence, France
{mohamed-aamine.ben-temim,guillaume.ferre,romain.tajan}@ims-bordeaux.fr

Abstract. The rapid growth in the Internet of Things (IoT) leads currently to an increase in the density of the number of connected objects typically in free bands. Operating in unlicensed bands requires connected objects to reduce their energy consumption. To that end, one of the adopted techniques is the random access to the radio channel. The main downside of uncontrolled medium access is almost always the high interference level inherent to both self- as well as cross-technologies. Among the various technologies that are used to connect this massive number of devices, we distinguish two emerging technologies based on spread spectrum and ultra narrow band. The goal of this paper is to analyze the impact of the coexistence of these two technologies, in Industrial, Scientific, and Medical (ISM) unlicensed bands, in terms of bit error rate by defining a formal expression for the interfering signals.

Keywords: IoT · Spread spectrum · Ultra narrow band · ISM bands · Collisions

1 Introduction

Currently the number of connected objects is growing rapidly. On the radio link, i.e. between the device and the gateway, free or licensed frequency bands can be used. Operating in free bandwidth, typically Industrial, Scientific, and Medical (ISM) bands, offers the possibility to different devices to access the spectrum and provide wide number of services as long they abide by regulations. The primary advantage is license cost effectiveness. Nevertheless, the main downside of uncontrolled channel access is the high interference levels.

The communication technologies used in the ISM bands are such that the used bands to communicate are either narrow-band or wide-band. Sigfox [1] is one of the most attractive IoT technologies that are based on Ultra Narrow-Band (UNB) as physical layer. While Sigfox's main competing technology, LoRa [2], is a spread spectrum technology. More precisely, it is the Chirp Spread Spectrum (CSS) technique [3] that is used to transmit information.

© Springer Nature Switzerland AG 2020
O. Habachi et al. (Eds.): UNet 2019, LNCS 12293, pp. 56–67, 2020.
https://doi.org/10.1007/978-3-030-58008-7_5

In the literature, several works have analyzed the impact of the interference between heterogeneous networks operating in ISM bands. For example, the impact on coverage and packet error rate is analyzed in [4], when networks based on CSS and UNB techniques coexist. This analysis is obtained from simulators and imposes exorbitant simulation times. The authors in [5] studied the effect of narrow-band interference on LoRa's decoding performance and highlighted the advantage of using an interference suppression technique. Finally, real measurements in [6] and [7] already investigate the impacts of interference in the 868 MHz ISM band on the coverage and capacity of two networks based on LoRa and Sigfox technologies.

The goal of this paper is to study the impact of the mutual interference between two networks that use UNB and spread spectrum technology on their radio interfaces and to evaluate their robustness. This study is carried out at the level of the physical layer. Indeed, we express and analyse the interference term disrupting the decision at the symbol time scale. To the best of our knowledge, there is no work that deals with collision evaluation through the analysis of the percent of the time overlap between UNB and CSS signals at the symbol time scale. In this context, our contribution consists in evaluating the signal decoding performance of these two techniques by providing theoretical models describing this phenomenon. On the basis of these models, we propose to evaluate numerically and at low complexity the mutual impact of the two communication techniques by observing the degradation of the bit error rate (BER) curves.

The remainder of this paper is organized as follows: In Sect. 2, we provide a brief overview of UNB and CSS technologies. Section 3 states the theoretical approach adopted to examine inter-interference impact on signal decoding performance. Before concluding our work, simulation results are proposed and commented.

2 Competing Technologies

This section is dedicated to present the two emerging Internet of Things (IoT) technologies CSS and UNB which are widely used in IoT systems. This choice is justified by the popularity of networks based on LoRa and Sigfox technologies.

2.1 Chirp Spread Spectrum Communications

Chirp Spread Spectrum is a spread spectrum technique that uses wideband linear frequency modulated chirp pulses to transmit the binary information. When LoRa is considered, the signal is transmitted over the entire allocated bandwidth using different spreading factors (SFs). Initially, the binary information flow generated from the MAC layer is divided into sub-sequences, each of length $SF \in [7...12]$ [8]. For a digital communication point of view, SF represents also the number of bits per symbol.

We define $f_c^m(t)$ as the chirp transmitted at time mT_s^{wb} [9], where T_s^{wb} is the CSS symbol duration. This chirp has been obtained using $\tau_m = \frac{S_m}{B_{wb}}$ and performing a cyclic shift as depicted by Fig. 1. It should be noted that S_m is a random value uniformly distributed in $[0, \ldots, 2^{SF} - 1]$, constituted from SF consecutive bits and corresponds to the transmitted symbol at time mT_s^{wb}. B_{wb} is the CSS signal bandwidth and $T_s^{wb} = \frac{2^{SF}}{B_{wb}}$ by definition [9].

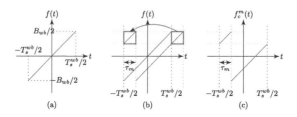

(a)   (b)   (c)

Fig. 1. Symbol $\rightarrow$ chirp association process - (a) up raw chirp - (b) process principle - (c) chirp associated to τ_m.

Given that $f_c^m(t)$ is the derivative of the phase $\phi_c^m(t)$:

$$f_c^m(t) = \frac{1}{2\pi} \frac{\mathrm{d}\phi_c^m(t)}{\mathrm{d}t} \tag{1}$$

Therefore, the instantaneous phase $\phi_c^m(t)$ can be expressed for $t \in \left[-\frac{T_s^{wb}}{2}, -\frac{T_s^{wb}}{2} + \tau_m\right)$:

$$\phi_c^m(t) = 2\pi \left[\frac{B_{wb}}{2T_s^{wb}} t^2 - \frac{S_m}{T_s^{wb}} t\right] \tag{2}$$

And for $t \in \left[-\frac{T_s^{wb}}{2} + \tau_m, \frac{T_s^{wb}}{2}\right)$:

$$\phi_c^m(t) = 2\pi \left[\frac{B_{wb}}{2T_s^{wb}} t^2 - \left(\frac{S_m}{T_s^{wb}} - B_{wb}\right) t\right] \tag{3}$$

If we note $s_{wb}(t)$ the complex envelope of the signal transmitted by a device based on CSS, we have:

$$s_{wb}(t) = \sum_{k \in \mathbb{Z}} e^{j\phi_c^k(t - kT_s^{wb})} \tag{4}$$

2.2   Ultra Narrow Band Communications

Ultra Narrow Band is usually referred to technology transmitting over very narrow spectrum channel, i.e. <1 kHz [10], to achieve low-rate and long communication distance (5 km in urban or 25 km+ in suburban area). This makes theoretical sense because of excellent link budget from low in-band receive noise. This type of bandwidth is particularly suited for small traffic with low bit rate, which makes it a good candidate for low-power wide area networks (LPWANs). Eventually, the expression of the baseband transmitted signal by one UNB device is given as follows:

$$s_{nb}(t) = \sum_{k \in \mathbb{Z}} A_k g(t - kT_s^{nb}) \tag{5}$$

where T_s^{nb} is the UNB symbol duration, A_k represents the *iid* transmitted symbol at time kT_s^{nb} and $g(t)$ is the basic pulse shaping filter.

The next section is dedicated to develop the theoretical models of the mutual interference between CSS and UNB signals, based on their complex envelop definitions.

3 Interference Model

The main problem with the coexistence of spread spectrum and ultra narrowband IoT networks in the ISM bands is the packet collisions. The objective of this section is to present theoretically the impact of UNB communications on a CSS signal and vice-versa.

To model our system, we present in Fig. 2 an illustrative scheme scaling UNB and CSS interference model.

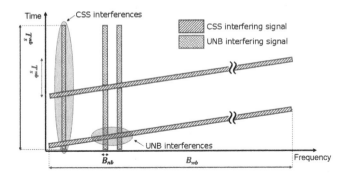

Fig. 2. Interference model between UNB and CSS symbols.

3.1 UNB Interfering Signals Impact on CSS Communication

In order to model the impact of UNB communication on the decoding of the spread spectrum signal, we consider a CSS signal in the presence of several UNB interferences. An uplink communication scenario is considered. Thus, the received signal at a gateway demodulating CSS signals, sampled at $T_e = \frac{T_s^{wb}}{N}$ with $N = 2^{SF}$ [8], is given by:

$$y^{wb}(nT_e) = \sqrt{P_{wb}} \sum_{k \in \mathbb{Z}} e^{j\phi_c^k(nT_e - kT_s^{wb})} + \sum_{i=1}^{N_{int}} \sqrt{P_{nb,i}} s_{nb,i}(nT_e) + w(nT_e) \tag{6}$$

where

- N_{int}: number of UNB interfering signals,
- P_{wb}: received power of the CSS signal,
- $P_{nb,i}$: received power of the ith UNB interfering signal,
- $w(nT_e)$: additive white gaussian noise (AWGN),
- $s_{nb,i}(nT_e)$: i^{th} UNB interfering signal.

In addition, by defining for the ith UNB interferer:

- $f_{p,i}$: the baseband carrier frequency,
- $\Delta\tau_i$: the time desynchronization,
- $A_{i,k'}$: the symbol transmitted at the time $k'T_s^{nb}$,

we obtain:

$$s_{nb,i}(nT_e) = \sum_{k'\in\mathbb{Z}} A_{i,k'}g(nT_e - k'T_s^{nb} - \Delta\tau_i)e^{j2\pi f_{p,i}nT_e} \tag{7}$$

Knowing the complex envelope of the raw chirp, the digital demodulation of the mth transmitted symbol is obtained during the following time interval: $mT_s^{wb} - \frac{T_s^{wb}}{2} \le t < mT_s^{wb} + \frac{T_s^{wb}}{2}$. If $r_m^{wb}(nT_e)$, $n \in [-\frac{N}{2}, \frac{N}{2} - 1]$ corresponds to the signal processed by the CSS demodulator, we have:

$$\begin{aligned} r_m^{wb}(nT_e) &= y^{wb}(nT_e + mT_s^{wb})e^{-j\phi_0(nT_e)} \\ &= s_m^{wb}(nT_e) + \sum_{i=1}^{N_{int}} i_{m,i}^{nb}(nT_e) + w_m(nT_e) \end{aligned} \tag{8}$$

where $\phi_0(nT_e) = 2\pi\frac{B_{wb}}{2T_s^{wb}}(nT_e)^2$ corresponds to the instantaneous phase of the raw chirp. In this case, the useful signal and the noise term are respectively equal to [9]:

$$s_m^{wb}(nT_e) = \sqrt{P_{wb}}e^{-j2\pi\frac{S_m n}{N}} \tag{9}$$

and

$$w_m(nT_e) = w(nT_e + mT_s^{wb})e^{-j\phi_0(nT_e)} \tag{10}$$

Whereas the ith UNB interfering signals, $i_{m,i}^{nb}(nT_e)$, is given by:

$$\begin{aligned} i_{m,i}^{nb}(nT_e) &= \sqrt{P_{nb,i}}s_{nb,i}(nT_e + mT_s^{wb} - \Delta\tau_i)e^{-j\phi_0(nT_e)} \\ &= \sqrt{P_{nb,i}}\sum_{k'\in\mathbb{Z}} A_{i,k'}g(nT_e + mT_s^{wb} - k'T_s^{nb} - \Delta\tau_i) \\ &\quad \times e^{j2\pi f_{p,i}(nT_e + mT_s^{wb})}e^{-j\phi_0(nT_e)} \end{aligned} \tag{11}$$

Here we notice that the receiver is perfectly synchronized on the received CSS signal and we suppose that the interfering signals do not affect this synchronization process.

The interference between a CSS symbol and a UNB signal occurred only when they overlap in the time and the frequency domains. This interference is characterized by the duration in which CSS symbol crosses the UNB bandwidth B_{nb}. Furthermore, the samples of $s_{nb,i}(nT_e + mT_s^{wb})$ that interfere with the mth CSS symbol are:

$$\Delta n_{int} = \{[n_1, n_2] \subset [-\frac{N}{2}, \frac{N}{2} - 1], \forall n \in [n_1, n_2],$$

$$f_c^m(nT_e) \in [f_{p,i} - \frac{B_{nb}}{2}, f_{p,i} + \frac{B_{nb}}{2}]\}$$

given that $f_c^m(nT_e)$ is an increasing function $\forall n \in [n_1, n_2]$. Thus, the number of interfering samples is given by:

$$n_{int} = \frac{B_{nb}2^{SF}}{B_{wb}^2 T_e} = \frac{B_{nb}T_s^{wb}}{B_{wb}} \frac{1}{T_e} \tag{12}$$

This prove that the number of interfering samples between CSS and UNB symbols increases by increasing the SF and the bandwidth of the UNB signal and decreases by increasing the bandwidth of the CSS signal. Then, the ith UNB interfering signal sampled at T_e, denoted by $i_{m,i}^{nb}(nT_e)$, with $n \in [-\frac{N}{2}, \frac{N}{2} - 1]$, is defined as:

$$i_{m,i}^{nb}(nT_e) = \begin{cases} zero & \text{for } n \in [-\frac{N}{2}, \frac{N}{2} - 1] \backslash [n_1, n_2] \\ (11) & \text{otherwise} \end{cases} \tag{13}$$

The optimal estimation of S_m and thus the detection of the associated symbol can be performed by searching for the maximum of $r_m(nT_e)$ periodogram [9]. If we note $R_m[k]$, $k \in [0, N-1]$, the Discrete Fourier Transform (DFT) of $r_m^{wb}(nT_e)$ we have:

$$R_m[k] = \frac{1}{\sqrt{N}} \sum_{n=-\frac{N}{2}}^{\frac{N}{2}-1} r_m^{wb}(nT_e)e^{-j2\pi\frac{nk}{N}} \tag{14}$$

By exploiting the DFT periodicity, $R_m[k]$ can be expressed as follows:

$$R_m[k] = R_m[k - N] = \sqrt{P_{wb}N}\delta(k + S_m - N) + I_m^{nb}[k - N] + W_m[k - N] \tag{15}$$

where $I_m^{nb}[k]$ and $W_m[k]$ are the DFT of the UNB interference and the noise respectively. The estimation of mth CSS symbol is then given by:

$$\hat{S}_m = N - \underset{k}{argmax}(|R_m[k]|^2) \tag{16}$$

The decision on the symbol S_m is perturbed by the noise and the interference terms. The interference term $I_m^{nb}[k]$ is the sum of the DFTs of the interfering signals defined in (13) which have the symbol interference duration as depicted in (12). We will use these expressions to numerically evaluate, with a low complexity, the UNB signal impact on the CSS symbol demodulation.

3.2   CSS Interfering Signals Impact on UNB Communication

In this section, we aim to analyze the impact of CSS signals on UNB communications. To do this, we consider the reception of an UNB signal in the presence of several CSS interfering signals.

Given that $y^{nb}(nT_e)$ corresponds to the received signal, sampled at T_e, at the gateway using UNB technology, we have:

$$y^{nb}(nT_e) = \sqrt{P_{nb}} \sum_{k' \in \mathbb{Z}} A_k g(nT_e - kT_s^{nb}) + \sum_{i=1}^{N_{int}} \sqrt{P_{wb,i}} s_{wb,i}(nT_e) + w(nT_e)$$

$$(17)$$

where N_{int} is the number of CSS interfering signals, P_{nb} (resp. $P_{wb,i}$) is the received power of UNB signal (resp. i^{th} CSS interfering signal), and $s_{wb,i}(nT_e)$ is the ith CSS interfering signal expressed as:

$$s_{wb,i}(nT_e) = e^{-j2\pi f_{nb}nT_e} \sum_{k \in \mathbb{Z}} e^{j\phi_{c,i}^k(nT_e - kT_{s,i}^{wb} - \Delta\tau_i)}$$

$$(18)$$

here f_{nb} represents the carrier frequency offset compensated by the UNB digital demodulator in order to process the set of UNB symbols to estimate and $\Delta\tau_i$ is the time desynchronization of the ith CSS interfering signal as we suppose that the receiver is perfectly synchronized on the UNB signal.

To maximize the signal to noise ratio (SNR), classical matched filter is performed at the receiver. If we note $g_a(nT_e)$ the nth sample of this filter, we obtain:

$$r^{nb}(nT_e) = y^{nb}(nT_e) * g_a(nT_e)$$

$$= \sqrt{P_{nb}} \sum_{k \in \mathbb{Z}} A_k v(nT_e - kT_s^{nb}) + \sum_{i=1}^{N_{int}} \sqrt{P_{wb,i}} i_i^{wb}(nT_e) + w'(nT_e) \quad (19)$$

where $w'(nT_e) = w(nT_e) * g_a(nT_e)$ (resp. $i_i^{wb}(nT_e) = s_{wb,i}(nT_e) * g_a(nT_e)$) is the filtered noise (resp. the filtered i^{th} CSS interfering signal) and $v(nT_e) = g(nT_e) * g_a(nT_e)$ is the Nyquist filter.

Then, the optimal estimation of the mth transmitted UNB symbol is obtained by sub-sampling at T_s^{nb} the matched filter output. Thus we have:

$$r^{nb}(mT_s^{nb}) = \sqrt{P_{nb}} A_m v(0) + i^{wb}(mT_s^{nb}) + w'(mT_s^{nb}) \quad (20)$$

The UNB symbol estimation is then obtained using classical digital demodulation processing.

If now we focus our attention on the CSS signal interfering the mth UNB symbol decision, we have:

$$i^{wb}(mT_s^{nb}) = \left(\sum_{i=1}^{N_{int}} \sqrt{P_{wb,i}} s_{wb,i}^m(nT_e) * g_a(nT_e) \right)_{mT_s^{nb}} \quad (21)$$

where

$$s_{wb,i}^m(nT_e) = e^{-j2\pi f_{nb}nT_e} \sum_{k \in \Omega_m^i} e^{j\phi_{c,i}^k(nT_e - kT_{s,i}^{wb} - \Delta\tau_i)} \quad (22)$$

Here Ω_m^i is the set of symbols of the ith CSS interfering signal that are eligible for the interference with the mth UNB symbol and we have:

$$\Omega_m^i = \{k \in \mathbb{Z}, -\frac{T_{s,i}^{wb}}{2} - \frac{T_s^{nb}}{2} < kT_{s,i}^{wb} - mT_s^{nb} < \frac{T_s^{nb}}{2} + \frac{T_{s,i}^{wb}}{2}\}.$$

The cardinality of Ω_m^i can be approximated as follows:

$$Card(\Omega_m^i) = \frac{T_s^{nb}}{T_{s,i}^{wb}} = \frac{B_{wb}}{2^{SF_i}B_{nb}} \tag{23}$$

Based on the interference duration between the CSS and UNB symbols presented in the previous study, we can define the samples of each symbol of $s_{wb,i}^m(nT_e)$ that affect the mth UNB symbol decision.

From the models of the interfering signals, (13) and (21), obtained in the previous section and from our observations, we propose in the next section to digitally analyse the mutual impact of CSS and UNB signals. Indeed, unlike a complete models of UNB and CSS systems, the later expressions allow to simulate interference phenomena with low complexity.

4 Simulations and Discussion

First of all, we choose to analyse the impact of the coexistence of CSS and UNB signals by observing the performance degradation on the bit error rate (BER). We consider that the digital modulation used by the UNB system is a differential binary phase shift keying (DBPSK) and that the signals are systematically in collision over their entire duration. The degradation related to every interference scenario is compared to the theoretical BER for CSS or DBPSK modulations without interference [4]. Figure 3 and 4 correspond to the use of the bandwidths adopted by LoRa and Sigfox ($B_{wb} = 125$ kHz and $B_{nb} = 100$ Hz). Both figures show that raising the power of interference results in decoding failure of the received signal. The results of Fig. 3 allow to quantify the robustness of a LoRa signal against an UNB interference (Sigfox type). Indeed, for a signal to interference ratio (SIR) $\gamma_l = 10\log_{10}(\frac{P_{wb}}{P_{nb}})$ of -20 dB we observe a loss of sensitivity of only 1 dB for a BER $= 10^{-4}$, when the $SF = 10$. Figure 4 allows to observe the robustness of the Sigfox technology to a spread spectrum interference (LoRa type). We can see that the smaller the SF of the interfering signal is, the more affected the Sigfox signals are. Indeed, as shown in (12) when the SF increases, although the duration of CSS interference on the time scale of the UNB symbol increases, its occurrence decreases considerably. By referring to (23), we can deduce for $SF = 7$ that 10 LoRa symbols affect the decision of one Sigfox symbol. However, for $SF = 12$ one LoRa symbol affect only one Sigfox symbol out of three.

Beyond these observations, Fig. 3 and 4, allow us to validate our interference models, since the results obtained with our Matlab simulators overlap with the

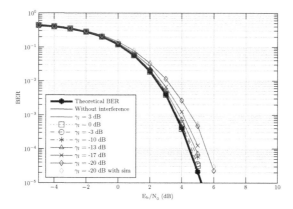

Fig. 3. Impact of a UNB signal on CSS communication - $SF = 10$, $B_{wb} = 125$ kHz, $B_{nb} = 100$ Hz and $\gamma_l = 10\log_{10}(\frac{P_{wb}}{P_{nb}})$.

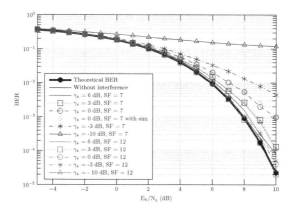

Fig. 4. Impact of a CSS signal on UNB communication - $B_{nb} = 100$ Hz, $B_{wb} = 125$ kHz and $\gamma_s = 10\log_{10}(\frac{P_{nb}}{P_{wb}})$.

results of our models. We illustrate this for the SIR $\gamma_l = -20$ dB on Fig. 3 and a $\gamma_s = 0$ dB with $SF = 7$ for Fig. 4.

In order to go further in the analysis of the results, we represent in Fig. 5 (resp. Fig. 6) the impact of UNB (resp. CSS) interference on the sensitivity of the CSS (resp. UNB) receiver. We note this sensitivity Γ_l for the CSS signal and Γ_s for the UNB one. The latter characterizes the floor value of $\frac{E_b}{N_0}$ to define the range of each technology. For LoRa-like signals, this sensitivity corresponds to a BER of 10^{-5} [11]. Thereafter, and without loss of generality, we limit this sensitivity to a BER of 10^{-4} for both technologies. Finally, considering that a signal remains UNB up to a bandwidth of 1 kHz, we show the impact of increasing the bandwidths of UNB and CSS signals on the receiver sensitivity. For the further results, we notice that $B = (B_{nb}, B_{wb})$ is the bandwidth configuration adopted for each interference scenario.

Figure 5 shows that the decoding quality of the CSS signal is clearly more degraded by increasing B_{nb}. This result is consistent with (12) which indicates that the duration of the interference between the CSS and UNB symbols is proportional to B_{nb}. In fact, for an SIR value equal to -10 dB, Γ_l increases by 1.5 dB (resp. 0.6 dB) compared to the reference sensitivity Γ_l^{ref} in the presence of an UNB interfering signal with B_{nb} equal to 1kHz (resp. 100 Hz). In addition, if we consider 10 and 40 UNB interfering signals, with carrier frequencies $f_{p,i}$ uniformly distributed in $[0, B_{wb}]$ and a bandwidth $B_{nb} = 100$ Hz, the sensitivity threshold increases slightly compared to the case of a single interfering signal with the same bandwidth. Finally, the green curves show that the decoding performance of the CSS signal is clearly enhanced when the CSS signal bandwidth is increased from 125 kHz to 250 kHz. This result is consistent with (12) which indicates that the duration of the interference between the CSS and UNB symbols is inversely proportional to B_{wb}.

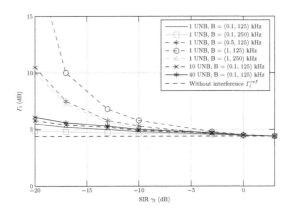

Fig. 5. Impact of UNB interference on CSS reception sensitivity Γ_l as function of $\gamma_l = 10\log_{10}(\frac{P_{wb}}{\sum_{i=1}^{N_{int}} P_{nb,i}}) - SF = 10$. (Color figure online)

Figure 6 confirms the fact that the decoding quality of the UNB signal is more affected by decreasing the SF of the interfering signal. For example, for an SIR of -3 dB and $B_{nb} = 100$ Hz, Γ_s increases by 6.9 dB (resp. 0.6 dB) compared to Γ_s^{ref} in the presence of a CSS interferer with a SF of 7 (resp. 12). Moreover, this figure shows that even if we increase B_{nb}, the performance degradation remains almost the same. This result is explained by the compensation of the increase in the duration of the interference on CSS symbols by a lower occurrence of this interference. Furthermore, if we consider the case of 10 CSS interfering signals with SFs uniformly distributed throughout the set $\{7, ..., 12\}$, we obtain, for an equivalent total interference power, a performance degradation less important than that in the presence of only one interfering signal with a $SF = 7$ and more

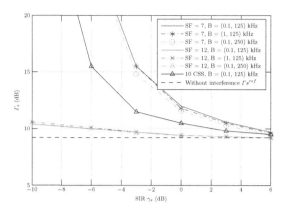

Fig. 6. Impact of CSS interference on UNB reception sensitivity Γ_s as function of $\gamma_s = 10 \log_{10}(\frac{P_{nb}}{\sum_{i=1}^{N_{int}} P_{wb,i}})$. (Color figure online)

important than the case when the SF of the later is equal to 12. Finally, the green curves show that increasing the CSS interfering signals bandwidth does not harm the signal decoding performance of the UNB signal. Given the expression of (12), the increase of B_{wb} leads to a decrease of the symbols interference duration, but leads also to an increase of the occurrence of this interference as given by (23). In this coexistence context, the following points can be deduced:

- CSS signals appear to be more resilient than UNB signals particularly for the lowest SFs of CSS interfering signals.
- Increasing UNB signals bandwidth does not lead to better robustness against the CSS interference. Furthermore, it degrades the decoding performance of the CSS signals.
- Increasing CSS signals bandwidth enhances clearly their decoding performance. However, it does not lead to better for the UNB case.

5   Conclusion

In this paper, we formally explained at the time symbol scale the impact of UNB interference on a CSS communication and vice versa. In this context, our contribution consisted in evaluating the signal decoding performance of these two techniques by providing theoretical models describing this phenomenon. On the basis of these expressions we have evaluated the mutual impact of the two communication techniques by observing the degradation of the BER and the sensitivity curves. This approach is original compared to what we find in the literature where the results are obtained using simulators which lead to significant simulation time. We validate our formal expressions of interference by corroborating them with results obtained on Matlab simulators developed in the laboratory.

Finally, in this coexistence context, our results show that the increase in the bandwidth of the UNB signals does not lead to a better robustness of the later to CSS interference. However, the increase in CSS signal bandwidth allow the later to be more resistant to UNB interference.

In terms of perspectives, we are quantifying the time savings in terms of execution time of our approach compared to the global simulator approach, then thanks to our interference models, we will be able to analyse realistic cases of deployment using realistic IoT networks.

Acknowledgement. This study has been carried out with financial support from the French State, managed by the French National Research Agency (ANR) in the frame of the "Investments for the future" Programme IdEx Bordeaux - SysNum (ANR-10-IDEX-03-02).

References

1. Sigfox: Sigfox - The Global Communications Service Provider for the Internet of Things (IoT) (2018). https://www.sigfox.com/en
2. Semtech: LoRa Modem Design Guide: Sx1272/3/6/7/8 (2013). https://www.semtech.com/uploads/documents/LoraDesignGuide_STD.pdf
3. Reynders, B., Pollin, S.: Chirp spread spectrum as a modulation technique for long range communication. In: 2016 Symposium on Communications and Vehicular Technologies (SCVT), pp. 1–5, November 2016
4. Reynders, B., Meert, W., Pollin, S.: Range and coexistence analysis of long range unlicensed communication. In: 23rd International Conference on Telecommunications (ICT), pp. 1–6, May 2016
5. Elshabrawy, T., Robert, J.: The impact of ISM interference on LoRa BER performance. In: 2018 IEEE Global Conference on Internet of Things, pp. 1–5, December 2018
6. Lauridsen, M., et al.: Interference measurements in the European 868 MHZ ISM band with focus on LoRa and SigFox. In: IEEE Wireless Communications and Networking Conference (WCNC), March 2017
7. Vejlgaard, B., et al.: Interference impact on coverage and capacity for low power wide area IoT networks. In: 2017 IEEE Wireless Communications and Networking Conference (WCNC), pp. 1–6, March 2017
8. Ferré, G., Giremus, A.: Lora physical layer principle and performance analysis. In: 2018 25th IEEE International Conference on Electronics, Circuits and Systems (ICECS), pp. 65–68, December 2018
9. Ferré, G., Simon, E.P.: Sigfox and LoRa PHY and MAC layers, Research Report, April 2018. https://hal.archives-ouvertes.fr/hal-01768341
10. Anteur, M., Deslandes, V., Thomas, N., Beylot, A.: Ultra narrow band technique for low power wide area communications. In: 2015 IEEE Global Communications Conference (GLOBECOM), pp. 1–6, December 2015
11. Elshabrawy, T., Robert, J.: Analysis of BER and coverage performance of LoRa modulation under same spreading factor interference. In: 2018 IEEE 29th Annual International Symposium on Personal, Indoor and Mobile Radio Communications (PIMRC), pp. 1–6, September 2018

Energy-Efficient MIMO Multiuser Systems: Nash Equilibrium Analysis

Hang Zou[1(✉)], Chao Zhang[1], Samson Lasaulce[1], Lucas Saludjian[2], and Patrick Panciatici[2]

[1] L2S, CNRS-CentraleSupelec-Univ. Paris Saclay, Gif-sur-Yvette, France
{hang.zou,chao.zhang,samson.lasaucle}@centralesupelec.fr
[2] RTE, Paris, France

Abstract. In this paper, an energy efficiency (EE) game in a MIMO multiple access channel (MAC) communication system is considered. The existence and the uniqueness of the Nash Equilibrium (NE) is affirmed. A bisection search algorithm is designed to find this unique NE. Despite being sub-optimal for deploying the ε-approximate NE of the game when the number of antennas in transmitter is unequal to receiver's, the policy found by the proposed algorithm is shown to be more efficient than the classical allocation techniques. Moreover, compared to the general algorithm based on fractional programming technique, our proposed algorithm is easier to implement. Simulation shows that even the policy found by proposed algorithm is not the NE of the game, the deviation w.r.t. to the exact NE is small and the resulted policy actually Pareto-dominates the unique NE of the game at least for 2-user situation.

Keywords: Energy efficiency · Multiple access channel · MIMO · Game theory · Nash Equilibrium · Approximate Nash Equilibrium

1 Introduction

With the release of first 5G package, it turns out that the number of devices in the upcoming wireless network will increase tremendously, e.g., Internet of Things (IoT). Consequently, classical paradigm which merely aims at optimizing the quantitative performance, e.g., data-rate, bit-error-rate and latency faces extreme difficulty in many domains in both academic research and industrial application. Thus the issue of energy-efficient design of the wireless system tends to be crucial. Different definition of energy efficiency (EE) has been proposed in recent years in [12–15]. Amongst which the most popular one is defined as the total benefit obtained under the unit consumption of energy or power known as global energy efficiency (GEE) e.g., in [3–5,11]. Taken the bits-per-second type rate function as benefit function, one will obtain the well-known bits-per-joule energy efficiency.

© Springer Nature Switzerland AG 2020
O. Habachi et al. (Eds.): UNet 2019, LNCS 12293, pp. 68–81, 2020.
https://doi.org/10.1007/978-3-030-58008-7_6

One of the pioneer works of studying the maximization of EE in Multiple-Input Multiple-Output (MIMO) system is [5]. In [5], the optimal precoding scheme is studied and divided into different cases with different assumptions on the systems. Till now the optimal precoding matrix for general condition is merely conjectured and unproved. Hereafter, optimal precoding matrix design for single user MIMO system is performed for imperfect channel state information (CSI) scenario in [9]. Then it is later widely realized that the problem of EE maximization actually belongs to the category of fractional programming. Techniques such as Dinkelbach's algorithm (see [8]) is used to solve EE maximization in [9,10]. These algorithms are generally based on the idea that the optimal solution can be found by solving a sequence of convex optimization problems related to the original one. The main difficulty of EE maximization OP is usually due to the non-convexity of energy efficiency function. Under some assumption on the benefit function, the EE function is well-known as being quasi-concave or even pseudo-concave. However, it is generally difficult to trace the Nash Equilibrium (NE) of a game where the individual utility function of player is of EE type. In [3], it is shown that there always exists an unique NE for scalar power allocation game in a relay-assisted MIMO systems due to the standard property of the best response dynamics. Similar results in MIMO-MAC system will be given latter in the paper.

The contribution of this paper is twofold: 1) we first extend the work in [3] to a more general situation where each user is allowed to choose its covariance matrix to maximize its individual EE instead of tuning its scalar power merely. The existence and uniqueness of the NE is proved under some assumptions. 2) An algorithm is proposed to find the unique NE of this MIMO-MAC game. When the number of antennas of transmitter is equal the one of receiver, proposed algorithm leads to exact NE. Otherwise it leads to the ε-approximate NE defined latter in the paper for replacing the exact best response dynamic by its linear approximation.

The remaining parts of the paper is organized as follows: the MIMO-MAC system and the EE game are first presented in Sect. 2. Then some basics of game theory are given and the existence and the uniqueness of NE of the EE game is proved in Sect. 3. In Sect. 4, a basic algorithm is proposed and an improved bisection search algorithm is given which yields an ε-approximate NE slightly Pareto-dominating the exact NE. The numeric results of proposed algorithms are compared with classical allocation policy and analyzed in Sect. 5. The paper concludes by several remarkable conclusions in Sect. 6.

Notations: $(\cdot)^H$ and $(\cdot)^\dagger$ denote matrix transpose and Moore-Penrose inverse respectively. $\mathbf{I}_N$ stands for identity matrix of size N. $\det(\cdot)$ and $\mathrm{Tr}(\cdot)$ denote the determinant and the trace of a matrix respectively. Denote the natural number set inferior or equal than N as $[N] \triangleq \{1, \ldots, N\}$.

2   System Model

Consider a multiple access channel (MAC) with one base station (BS) and K users (players) to be served. BS is equipped with N_r receive antennas and each user terminal is equipped with N_t transmit antennas. We assume a block fading channel where the realization of channel remains a constant during the coherence time of transmission and randomly generated according to some statistical distributions from period to period. The received signal at BS is given by:

$$y = \sum_{k=1}^{K} \mathbf{H}_k \boldsymbol{x}_k + \boldsymbol{z}, \tag{1}$$

where $\mathbf{H}_k \triangleq [\mathbf{H}_{k,i,j}]_{i,j=1}^{N_r,N_t} \in \mathbb{C}^{N_r \times N_t}$ is the channel transmit matrix of k-th user and $\mathbf{H}_{k,i,j}$ is the channel from i-th transmit antenna of k-th user to j-th receive antenna at BS which is assumed to be i.i.d. complex Gaussian distributed according to $\mathcal{CN}(0,1)$. $\boldsymbol{x}_k = (x_{k,1}, \ldots, x_{k,N_t})^T$ is the transmit symbol of k-th user and $\boldsymbol{z}$ is the noise observed by the receiver with complex Gaussian distribution $\mathcal{CN}(\mathbf{0}, \sigma^2 \mathbf{I}_{N_r})$. For the sake of simplicity, we assume that single user decoding is implemented for each user. Then the capacity the k-user can be achieved is

$$R_k = \log \frac{\det \left(\sigma^2 \mathbf{I}_{N_r} + \sum_{j=1}^{K} \mathbf{H}_j \mathbf{Q}_j \mathbf{H}_j^H \right)}{\det \left(\sigma^2 \mathbf{I}_{N_r} + \sum_{j \neq k}^{K} \mathbf{H}_j \mathbf{Q}_j \mathbf{H}_j^H \right)}, \tag{2}$$

where $\mathbf{Q}_k = \mathbb{E}[\boldsymbol{x}_k \boldsymbol{x}_k^H] \in \mathbb{C}^{N_t \times N_t}$ is the covariance matrix of symbol $\boldsymbol{x}_k$ which determines how power should be allocated for each antenna and $P_c > 0$ is the power dissipated in transmitter circuit to operate the devices. It is reasonable to assume that each user has perfect knowledge about its own channel, e.g., through downlink pilot training. Therefore user k is able to perform the singular value decomposition (SVD) of its own channel $\mathbf{H}_k$ and its covariance matrix $\mathbf{Q}_k$ as well. The SVD of $\mathbf{H}_k$ and $\mathbf{Q}_k$ is given by $\mathbf{H}_k = \mathbf{U}_k \mathbf{\Lambda}_k \mathbf{V}_k^H$ and $\mathbf{Q}_k = \mathbf{W}_k \mathbf{P}_k \mathbf{W}_k^H$ respectively. To simplify the problem, we assume that user k always adapts its covariance matrix to $\mathbf{H}_k$, i.e., choosing $\mathbf{W}_k = \mathbf{V}_k$. $\mathbf{P}_k$ is a diagonal matrix with $\mathbf{P}_k = \text{diag}(\boldsymbol{p}_k) = \text{diag}(p_{k1}, \ldots, p_{kN_t})$ where we use $\text{diag}(\cdot)$ to generate a diagonal matrix from a vector or vice versa. Thus user k's only legal action is represented by $\boldsymbol{p}_k$ or $\mathbf{P}_k$ and the action set of k-th user is

$$\mathcal{P}_k = \left\{ \boldsymbol{p}_k \left| \sum_{i=1}^{N_t} p_{ki} \leq \overline{P}_k, \ p_{ki} \geq 0 \right. \right\} \tag{3}$$

where $\overline{P}_k$ is power budget of k-th user. Through out the paper, we will use the matrix $\mathbf{P}_k$ or its diagonal $\boldsymbol{p}_k$ interchangeably to represent user k's action depending on the context. Further more, we denote $\boldsymbol{p} = (\boldsymbol{p}_k, \boldsymbol{p}_{-k})$ with $\boldsymbol{p}_{-k} \triangleq (\boldsymbol{p}_1, \ldots, \boldsymbol{p}_{k-1}, \boldsymbol{p}_{k+1} \ldots, \boldsymbol{p}_K) \in \mathcal{P}_{-k}$ and $\mathcal{P}_{-k} \triangleq \mathcal{P}_1 \times \cdots \times \mathcal{P}_{k-1} \times \mathcal{P}_{k+1} \times \cdots \times \mathcal{P}_K$. In this paper, energy efficiency defined as the ratio of a benefit function over the power consumed by producing it can be proven to has the following expression for user k after some simplifications:

$$u_k\left(\mathbf{P}_k,\mathbf{P}_{-k}\right) = \frac{\log \frac{\det\left(\sigma^2\mathbf{I}_{N_r}+\sum_{j=1}^K \mathbf{U}_j\mathbf{\Lambda}_j\mathbf{P}_j\mathbf{\Lambda}_j^H\mathbf{U}_j^H\right)}{\det\left(\sigma^2\mathbf{I}_{N_r}+\sum_{j\neq k}^K \mathbf{U}_j\mathbf{\Lambda}_j\mathbf{P}_j\mathbf{\Lambda}_j^H\mathbf{U}_j^H\right)}}{\mathrm{Tr}\left(\mathbf{P}_k\right)+P_c} \tag{4}$$

To this end, the MIMO MAC EE game is thus given by the following strategic form in triplet:

$$\mathcal{G} = \left(\mathcal{K},\left(\mathcal{P}_k\right)_{k\in\mathcal{K}},\left(u_k\right)_{k\in\mathcal{K}}\right) \tag{5}$$

3   Game-Theoretic Analysis

In this section, we will firstly give some basic concepts of any game-theoretic analysis. The central concept of game-theoretic analysis is Nash Equilibrium (NE) defined as:

Definition 1. *For game* $\mathcal{G} = \left(\mathcal{K},\left(\mathcal{P}_k\right)_{k\in\mathcal{K}},\left(u_k\right)_{k\in\mathcal{K}}\right)$, *an action profile* $\boldsymbol{p} = \left(\boldsymbol{p}_k,\boldsymbol{p}_{-k}\right)$ *is called a Nash Equilibrium if for* $\forall k \in \mathcal{K}$ *and* $\forall \boldsymbol{p}' = \left(\boldsymbol{p}_k',\boldsymbol{p}_{-k}\right)$:

$$u_k\left(\boldsymbol{p}_k,\boldsymbol{p}_{-k}\right) \geq u_k\left(\boldsymbol{p}_k',\boldsymbol{p}_{-k}\right) \tag{6}$$

The meaning of NE is that any unilateral change of action at this point won't lead to an enhance of individual benefit. Furthermore, we introduce an important conception in game-theoretic analysis known as best response dynamics.

Definition 2. *(Best Response): In a non-cooperative game* $\mathcal{G}$, *the correspondence* $\mathrm{BR}_k\left(\boldsymbol{p}_{-k}\right)$: $\mathcal{P}_{-k} \to \mathcal{P}_k$ *s.t.*

$$\mathrm{BR}_k\left(\boldsymbol{p}_{-k}\right) \triangleq \arg\max_{\boldsymbol{p}_k\in\mathcal{P}_k} u_k\left(\boldsymbol{p}_k,\boldsymbol{p}_{-k}\right) \tag{7}$$

is called the best response (BR) of player $k \in \mathcal{K}$ *given the action profile of other player* $\boldsymbol{p}_{-k}$. *From the definition of best response, one has immediately the following characterization for NE:*

Proposition 1. *[Nash,1950] An action profile* $\boldsymbol{p}^\star$ *is an NE if and only if:* $\forall k \in \mathcal{K}$, $\boldsymbol{p}_k^\star \in \mathrm{BR}_k\left(\boldsymbol{p}_{-k}^\star\right)$.

To identify the NE of game in (5), the properties of individual utility function should be identified as first step. We define two critical properties satisfied by the individual utility function.

Definition 3. *(Quasi-concavity) Let* $\mathcal{X} \in \mathbb{R}^n$ *be a convex set, a function* $f : \mathcal{X} \to \mathbb{R}$ *is said to be quasi-concave if*

$$f\left(\lambda\boldsymbol{x} + (1-\lambda)\boldsymbol{y}\right) \geq \min\left\{f\left(\boldsymbol{x}\right),f\left(\boldsymbol{y}\right)\right\} \tag{8}$$

for any $\boldsymbol{x},\boldsymbol{y} \in \mathcal{X}$ *with* $\boldsymbol{x} \neq \boldsymbol{y}$ *and* $0 < \lambda < 1$.

Definition 4. *(Pseudo-concavity) Let* $\mathcal{X} \in \mathbb{R}^n$ *be a convex set, a function* f : $\mathcal{X} \to \mathbb{R}$ *is said to be quasi-concave if it is differentiable and for any* $\boldsymbol{x}, \boldsymbol{y} \in \mathcal{X}$, *it holds:*

$$f(\boldsymbol{y}) < f(\boldsymbol{x}) \implies \nabla f(\boldsymbol{y})^T (\boldsymbol{x} - \boldsymbol{y}) > 0 \tag{9}$$

With the definition of quasi-concavity and the pseudo-concavity, Proposition 2 shows that the individual utility function does possess these important properties:

Proposition 2. R_k *is a concave functions w.r.t.* $\boldsymbol{p}_k$ *and* u_k *is a pseudo-concave (quasi-concave) function w.r.t.* $\boldsymbol{p}_k$ *for* $\forall k \in \mathcal{K}$; *For any fixed* $\boldsymbol{p}_{-k} \in \mathcal{P}_{-k}$ *and* p_{kj} *with* $j \neq i$, *only one of following statements is true for all* $i \in [N_t]$:

i) $\exists p_{ki}^\star > 0$ *s.t.* u_k *is an increasing function in* $(0, p_{ki}^\star)$ *and a decreasing function in* $(p_{ki}^\star, +\infty)$ *w.r.t.* p_{ki}.
ii) u_k *is a decreasing function in* $(0, +\infty)$ *w.r.t.* p_{ki}.

Proof. It is well-known that R_k is a concave function for $\boldsymbol{p}_k$. Then the pseudo-concavity (quasi-concavity) of u_k comes from the fact that it is a ratio of a concave function and an affine function of $\boldsymbol{p}_k$. For more details of the proof, see [2]. Now we prove the second part of this proposition. Rewrite the individual utility function as $u_k = \frac{R_k(\gamma_k)}{\sum_{i=1}^{N_t} p_{ki} + P_c}$ with $R_k(\gamma_k) = \log(1 + \gamma_k)$. Then we can prove that $\frac{\partial^2 u_k}{\partial p_{ki}^2} \leq 0$ due to the fact that R_k is an increasing concave function w.r.t. γ_k and γ_k is a also increasing concave function w.r.t. p_{ki}. However we can't conclude directly of the sign of $\lim_{p_{ki} \to +\infty} \frac{\partial u_k}{\partial p_{ki}}$. It can be positive or negative depending on the value of p_{kj} with $j \neq i$. Therefore, if $\lim_{p_{ki} \to +\infty} \frac{\partial u_k}{\partial p_{ki}} \geq 0$ then we are in case ii), otherwise we are in case i).

Before stating the best response dynamics of the game, we define the following boundary of set $\mathcal{P}_k$ indicated by an index subset $\mathcal{E} \subset [N_t]$:

$$\mathcal{P}_k[\mathcal{E}] \triangleq \{\boldsymbol{p}_k \in \mathcal{P}_k, p_{ki} = 0 \text{ for } i \in \mathcal{E}\} \tag{10}$$

and the non-negative index set for a given action $\mathbf{P}_k$:

$$\mathcal{I}(\mathbf{P}_k) \triangleq \{i \in [N_t] \text{ s.t. } p_{ki} \geq 0\} \tag{11}$$

Proposition 3. *For any given* $\mathbf{P}_{-k}$ *and provided that the power budget* $\overline{P}_k$ *is sufficiently large, denote the unique solution of the following equation as* $\mathbf{P}_k^*$:

$$diag\left(\boldsymbol{\Lambda}_k^H \left(\boldsymbol{\Lambda}_k \mathbf{P}_k \boldsymbol{\Lambda}_k^H + \mathbf{F}_k + \sigma^2 \mathbf{I}_r\right)^{-1} \boldsymbol{\Lambda}_k\right) = u_k(\mathbf{P}_k, \mathbf{P}_{-k}) \mathbf{I}_{N_t} \tag{12}$$

with $\mathbf{F}_k = \sum_{j \neq k} \mathbf{S}_j \mathbf{P}_j \mathbf{S}_j^H$ *is the interference matrix of k-th user with* $\mathbf{S}_j = \mathbf{U}_k^H \mathbf{U}_j \boldsymbol{\Lambda}_j$. *Then the BR of* $\mathbf{P}_k$ *w.r.t.* $\mathbf{P}_{-k}$ *is standard and converges to the unique NE admitted by game (5); The BR is the unique solution of (12) restricted to the boundary of* $\mathcal{P}_k$ *indicated by* $\mathcal{I}(\mathbf{P}_k^*)$ *with* $\mathcal{I}(\mathbf{P}_k^*) \neq \emptyset$.

Proof. our proof consists of two parts: i) existence of NE; ii) uniqueness of NE. i) Existence of NE: it is easy to prove that the action set $\mathcal{P}_k$ for each player is compact (closed and bounded), combining the quasi-concavity of u_k claimed in Proposition 2, the existence is due to Debreu-Fan-Glicksberg theorem [7]. Moreover, Proposition 2 claims that u_k is a pseudo-concave function w.r.t. $\mathbf{P}_k$. Due to the property of pseudo-concave function, the unique stationary point (points where derivative vanishes) is the global optimizer of the utility function if the stationary point is in the feasible action set. We first calculate the stationary point of u_k for $\forall k \in \mathcal{K}$ using matrix calculus which leads to (12). However, the stationary point might not belong to the feasible action set $\mathcal{P}_k$. Denote $\mathbf{P}_k^*$ the unique solution of (12) in $\mathbb{R}^{N_t}$. It is easy to prove that for given $\mathbf{P}_{-k}$, p_{ki}^* is a decreasing function w.r.t. $\forall p_{kj}$ with $j \neq i$ by contradiction. Due to this monotonicity of the BR and knowing that the feasible action set $\mathcal{P}_k$ is a polyhedron, BR must be on the boundary of $\mathcal{P}_k$ except $\mathbf{0}_{N_t \times N_t}$ defined as (10) which corresponds to the index set $\mathcal{I}(p_k^*) \neq \emptyset$, which completes the proof for existence.

ii) Now we would like to prove that the BR converges to a point which is the unique NE of the game. We will achieve that by showing that the best response is a standard function[1], i.e.,

1) Positivity: $\forall \mathbf{P}_{-k} \succcurlyeq 0$, $\mathrm{BR}_k(\mathbf{P}_{-k}) \succcurlyeq 0$;
2) Monotonicity: if $\mathbf{P}'_{-k} \succcurlyeq \mathbf{P}_{-k}$, then $\mathrm{BR}_k\left(\mathbf{P}'_{-k}\right) \succcurlyeq \mathrm{BR}_k(\mathbf{P}_{-k})$;
3) Scalability: $\mathrm{BR}_k(\alpha \mathbf{P}_{-k}) \prec \alpha \mathrm{BR}_k(\mathbf{P}_{-k})$ for any $\alpha > 1$.

Positivity is obviously observed in its form given by Proposition 3. The proof for monotonicity and scalability is similar to [3]. The strict proof is omitted due to the limit of space.

4   Algorithm for Finding NE

Proposition 3 actually provides an approach for us to find the NE of the game (5). One can easily deduce an iterative equation according to (12):

$$\mathrm{diag}\left(\mathbf{\Lambda}_k^H \left(\mathbf{\Lambda}_k \mathbf{P}_k^{(t)} \mathbf{\Lambda}_k^H + \mathbf{F}_k^{(t-1)} + \sigma^2 \mathbf{I}_r\right)^{-1} \mathbf{\Lambda}_k\right) = u_k\left(\mathbf{P}_k^{(t-1)}, \mathbf{P}_{-k}^{(t-1)}\right) \mathbf{I}_{N_t} \tag{13}$$

However, due to Proposition 3, this stationary point might not be in the feasible action set. One can design the following basic algorithm to find NE of the game (5) based on Proposition 3 summarized in Algorithm 1.

---

[1] The generalized inequality for matrix defined here is referred to its diagonal and takes the non-negative orthant as the underlying cone.

Algorithm 1. Basic Algorithm for finding NE of MIMO-MAC EE game

---

Initialization: $\mathbf{P}_k^{(0)} = \frac{1}{N_t}\mathbf{I}_{N_t}, \forall k$. Choose T and ϵ
For $t = 1$ to T, **do**
  For $k = 1$ to K, **do**
    Compute $\mathbf{P}_k^{(t)}$ using (13)
    If $\mathcal{I}\left(\mathbf{P}_k^{(t)}\right) \neq [N_t]$
      Compute $\mathbf{P}_k^{(t)}$ using (13) restricted to $\mathcal{I}\left(\mathbf{P}_k^{(t)}\right)$
    End If
  End For
  If $\sum_k \left\|\mathbf{P}_k^{(t)} - \mathbf{P}_k^{(t-1)}\right\| < \epsilon$
    Break
  End If
End For
Output: $\mathbf{P}_k^{\mathrm{NE}} = \mathbf{P}_k^{(t)}$ for $\forall k$.

Nevertheless, Algorithm 1 is not satisfactory way to find the NE of the game. More precisely, to find the BR for given $\mathbf{P}_{-k}$, one actually need to solve an optimization problem. However, if $h = U\left(\mathbf{P}_{-k}\right) = \max_{\mathbf{P}_k \in \mathcal{P}_k} u_k\left(\mathbf{P}_k, \mathbf{P}_{-k}\right)$ is known as *a priori* information, (13) can be transformed into following equation which is relatively easy to be solved compared to (13):

$$\mathrm{diag}\left(\mathbf{\Lambda}_k^H\left(\mathbf{\Lambda}_k\mathbf{P}_k^{(t)}\mathbf{\Lambda}_k^H + \mathbf{F}_k^{(t-1)} + \sigma^2\mathbf{I}_r\right)^{-1}\mathbf{\Lambda}_k\right) = h\mathbf{I}_{N_t} \tag{14}$$

Introducing an auxiliary parameter h, one obtains an iterative equation of $\mathbf{P}_k$. Without loss of generality, we assume that the solution of (13) belongs to the feasible action set for given $\mathbf{P}_{-k}$. Otherwise, similar analysis can applied for $\mathbf{P}_k$ but restricted on a boundary given by Proposition 3. For the sake of simplicity, we omit the discussion here and restrict ourselves to the situation where the BR is strictly included in the interior of the feasible action set. Therefore for all $i \in [N_t]$, there exists $p_{ki}^\star$ such that individual utility function $u_k\left(\mathbf{P}_k, \mathbf{P}_{-k}\right)$ is an increasing function in $(0, p_{ki}^\star)$ and a decreasing function in $(p_{ki}^\star, +\infty)$ with respect to p_{ki}, where $p_{ki}^\star$ is the i-th component of user k's BR for given $\mathbf{P}_{-k}$. Then u_k is also an increasing function in $(0, U\left(\mathbf{P}_{-k}\right))$ and a decreasing function in $(U\left(\mathbf{P}_{-k}\right), +\infty)$ w.r.t. parameter h. In other words, to find $\mathbf{P}_k = \mathrm{BR}\left(\mathbf{P}_{-k}\right)$, it is sufficient to find $U\left(\mathbf{P}_{-k}\right)$ by a bisection search due to the special monotonicity of the utility function.

However, it is worth mentioning that it is still difficult to directly find the solution of iterative Eq. (14). Because this solution is actually implicitly given. We would like to further simplify (14) to facilitate the calculation of BR or NE.

To start with, we assume that $N_t = N_r$. Firstly, we remove the diagonal operator of LHS of (14). Therefore we have:

$$\mathbf{P}_k^{(t)} = \frac{1}{h}\mathbf{I}_{N_t} - \mathbf{\Lambda}_k^{-1}\left(\mathbf{F}_k^{(t-1)} + \sigma^2\mathbf{I}_{N_r}\right)\mathbf{\Lambda}_k^{-1} \tag{15}$$

If $N_t > N_r$ or $N_t < N_r$ then $\mathbf{\Lambda}_k$ is not directly invertible, then we should consider the pseudo-inverse matrix of $\mathbf{\Lambda}_k$. Without loss of generality, we assume that $N_t > N_r$, denoting the right pseudo-inverse of $\mathbf{\Lambda}_k$ as $\mathbf{\Lambda}_k^\dagger$ then one has $\mathbf{\Lambda}_k\mathbf{\Lambda}_k^\dagger = \mathbf{I}_{N_r}$ and $\left(\mathbf{\Lambda}_k^\dagger\right)^H\mathbf{\Lambda}_k^H = \mathbf{I}_{N_r}$. Similarly, one has:

$$\mathbf{\Lambda}_k^H\left(\mathbf{\Lambda}_k\mathbf{P}_k^{(t)}\mathbf{\Lambda}_k^H + \mathbf{F}_k^{(t-1)} + \sigma^2\mathbf{I}_r\right)^{-1}\mathbf{\Lambda}_k = h\mathbf{I}_{N_t}$$

$$\left(\mathbf{\Lambda}_k\mathbf{P}_k^{(t)}\mathbf{\Lambda}_k^H + \mathbf{F}_k^{(t-1)} + \sigma^2\mathbf{I}_r\right)^{-1} = h\left(\mathbf{\Lambda}_k^\dagger\right)^H\mathbf{\Lambda}_k^\dagger \tag{16}$$

However, it is generally impossible to have $\mathbf{\Lambda}_k^\dagger\mathbf{\Lambda}_k = \mathbf{I}_{N_t}$. Thus the equality does not always holds when we multiply $\mathbf{\Lambda}_k^\dagger$ on left and $\left(\mathbf{\Lambda}_k^\dagger\right)^H$ on the right on both sides of the equation. Nevertheless, this operation will yield a linear approximation of the BR dynamics:

$$\widehat{\mathbf{P}}_k^{(t)} = \frac{\mathbf{\Lambda}_k^\dagger\left[\left(\mathbf{\Lambda}_k^\dagger\right)^H\mathbf{\Lambda}_k^\dagger\right]^{-1}\left(\mathbf{\Lambda}_k^\dagger\right)^H}{h} - \mathbf{\Lambda}_k^\dagger\left(\mathbf{F}_k^{(t-1)} + \sigma^2\mathbf{I}_{N_r}\right)\left(\mathbf{\Lambda}_k^\dagger\right)^H \tag{17}$$

Similarly, if $N_t < N_r$ we can obtain exactly the same iterative equation as (17). This type of dynamics belongs to the so-called ε-approximate best response which generally leads to the ε-approximate Nash Equilibrium defined as:

Definition 5. *For game* $\mathcal{G} = \left(\mathcal{K}, (\mathcal{P}_k)_{k\in\mathcal{K}}, (u_k)_{k\in\mathcal{K}}\right)$, *an action profile* $\boldsymbol{p} = \left(\boldsymbol{p}_k, \boldsymbol{p}_{-k}\right)$ *is called an* ε-*approximate Nash Equilibrium if for* $\forall k \in \mathcal{K}$ *and* $\boldsymbol{p}' = \left(\boldsymbol{p}_k', \boldsymbol{p}_{-k}\right)$ *for* $\varepsilon \geq 0$:

$$u_k\left(\boldsymbol{p}_k, \boldsymbol{p}_{-k}\right) - u_k\left(\boldsymbol{p}_k', \boldsymbol{p}_{-k}\right) \geq -\varepsilon \tag{18}$$

Obviously ε-approximate Nash Equilibrium is actually an extension of the concept of Nash Equilibrium. Notice that when $\varepsilon = 0$ then we are exactly back to the definition of Nash Equilibrium. If one deploys (17) as the BR dynamics to compute NE according to Algorithm 2, one may only result in ε-approximate Nash Equilibrium of the game. To this end, we obtain a sub-optimal algorithm summarized in Algorithm 2 by using the iterative equation deduced in (17) instead of using (13).

Algorithm 2. Bisection Search Algorithm for find the NE of MIMO-MAC EE game

---

Initialization: $\mathbf{P}_k^{(0)} = \frac{1}{N_t}\mathbf{I}_{N_t}, \forall k$. choose T, ϵ_1 and ϵ_2
For $t = 1$ to T, **do**
  For $k = 1$ to K, **do**
    Initialization: $\underline{h} = 0$ and $\overline{h} = h_{max}$
    Repeat Until $\overline{h} - \underline{h} \leq \epsilon_1$
      $h_M = \frac{\underline{h}+\overline{h}}{2}$, $h_L = \max\left(0, h_M - \frac{\epsilon_1}{2}\right)$ and $h_R = \min\left(h_{max}, h_M + \frac{\epsilon_1}{2}\right)$
      Compute $\mathbf{P}_k(h_i)$ using (17), $i \in \{L, M, R\}$
      $U_i = u_k\left(\mathbf{P}_k(h_i), \mathbf{P}_{-k}^{(t-1)}\right)$, $i \in \{L, M, R\}$
      If $U_L < U_M < U_R$
        $\underline{h} = h_L$
        Else If $U_L > U_M > U_R$
          $\overline{h} = h_R$
        End If
      Else
        $\underline{h} = h_L$ and $\overline{h} = h_R$
      End If
      Compute $\mathbf{P}_k^{(t)}$ by (17) with $h = h_M$
  End For
  If $\sum_k \left\|\mathbf{P}_k^{(t)} - \mathbf{P}_k^{(t-1)}\right\| < \epsilon_2$
    Break
  End If
End For
Output: $\mathbf{P}_k^{NE} = \mathbf{P}_k^{(t)}$ for $\forall k$.

Remark 1. Algorithm 2 actually works for general utility function possessing the same property as u_k. Moreover, this algorithm should be slightly faster than general bisection search. The reason is once by coincidence that the case neither $U_L > U_M > U_R$ nor $U_L < U_M < U_R$ occurs, we are surely to be very close to the stationary point of function. Otherwise we are in the monotonic region of the function, then this algorithm works as regular bisection search algorithm. In the worst case, this algorithm should have same complexity as the general bisection algorithm. Finally, Algorithm 2 merely requires the value of utility function instead of derivative of the utility function. Notice that Algorithm 2 is actually an off-line learning algorithm. Therefore a online-learning-version of Algorithm 2 by combing it with some deep learning techniques could be an important extension of this paper.

5   Numeric Results

The goal of this part is to show the performance of the proposed algorithms. Notice if $N_t = N_r$, (17) degenerates to (15) which conserves the optimality of best response. For this situation, we choose $N_t = N_r = 2$ with $K = 2$ users. A sufficient large power budget is chosen so that the BR is included in the feasible action set $\overline{P}_k = 10\,\mathrm{mW}$ for $\forall k \in \{1, 2\}$ and the circuit power is $P_c = 1\,\mathrm{mW}$. The error tolerance for Algorithm 2 is $\epsilon_1 = \epsilon_2 = 0.001$.

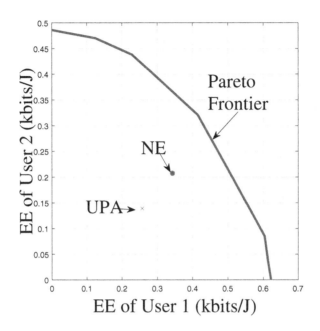

Fig. 1. Energy Efficiency under NE and uniform power allocation with $N_t = N_r = 2$ for 2-user situation. Policy found by our algorithms outperforms UPA policy.

In Fig. 1, the achievable utility region, the average performance under NE found by Algorithm 2 and the averaged performance achieved by uniform power allocation (UPA) are depicted. All results are averaged over 1000 randomly generated channel samples. It is observed that the performance achieved by deploying UPA is Pareto-dominated by NE which can be found by Algorithm 2. Furthermore, the NE found by Algorithm 2 is closed to the Pareto frontier achieved by some centralized algorithms which suggest the efficiency using Algorithm 2 is higher than UPA.

Moreover, define the social welfare for a given action profile as $w(\boldsymbol{p}) = \sum_{k \in \mathcal{K}} u_k(\boldsymbol{p}_k, \boldsymbol{p}_{-k})$. Then the average social welfare as function of number of number of antennas (still we keep $N_t = N_r$) and the power budget in Fig. 2 and Fig. 3 respectively. For Fig. 2, the averaged social welfare of both UPA policy and our proposed algorithm is increased quasi-linearly as the number of antennas grows. However our proposed algorithm always outperforms the optimal UPA policy which is allowed to tune the power but always equally shared among each transmit antenna. In Fig. 3, we would like to show the influence of user's power budget. There are two different regions for social welfare. In the first region where the power budget is sufficiently large, the NE found by our

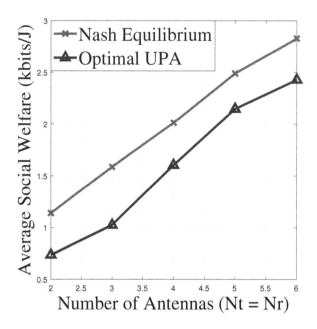

Fig. 2. Average social welfare under NE and UPA as function of number of antennas $(N_t = N_r)$ with $\overline{P}_k = 10\,\text{mW}$ for 2-user situation.

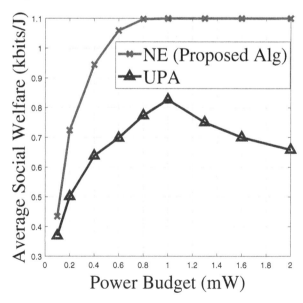

Fig. 3. Performance under NE and UPA as function of the power budget of user with $N_t = N_r = 2$ for 2-user situation. There are two different regions: one corresponds to Proposition 3. In the region uncovered by Proposition 3, proposed algorithm still dominates UPA.

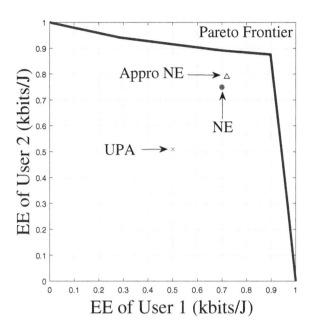

Fig. 4. Performance achieved by Algorithm 1 (NE) and Algorithm 2 (Approximate NE) and UPA with $N_t = 2$ and $N_r = 4$ for 2-user situation. Policy found by Algorithm 2 is very near to the exact NE and Pareto-dominates it. Moreover, two policies found by proposed algorithms both outperform UPA.

proposed algorithm is independent of the power budget while the performance of UPA is decreasing with respect to the increase of the power budget. In the second region where the power budget is relatively small, Using proposed algorithm, it is not sure to converge to the NE of the game because Proposition 3 is no more valid in this region. Nevertheless, the performance achieved by our algorithm is still better than UPA which prove the superiority of our algorithm.

Then a more probable situation is considered where $N_t < N_r$ meaning that the number of antennas in user terminal is less than the one in base station. The discussion in Sect. 4 shows that the proposed suboptimal algorithm is actually suboptimal due to the usage of ε-approximate best response. For numeric demonstration, we choose $N_t = 2 < N_r = 4$. The performance of Algorithm 2 is illustrated in Fig. 4. The sub-optimality is clearly demonstrated in this figure. However, the resulted policy actually Pareto-dominates the exact NE found by Algorithm 1 and the dispersion is relatively small in terms of average performance. This remark entails that even the policy found by Algorithm 2 is not the NE of the game in its sub-optimal region however its performance does slightly outperforms the exact NE. Moreover the proposed algorithm is easy to implement for using explicit iterative equation even if it is approximated.

6 Conclusions

In this paper, a game where the individual utility function is the energy efficiency in a MIMO multiple access channel system is considered. The existence and the uniqueness of Nash Equilibrium is proved and an exact algorithm and a suboptimal algorithm is proposed to find the NE of this game. Simulation results show that if the number of transmit antennas and the number of receiving antennas is the same, performance under NE found by proposed algorithms is always better than uniform power allocation policy for both inside or outside the range covered by the main proposition of the paper. When the condition for antennas is not met, our proposed algorithm actually deploys an ε-approximate best response which might leads to an ε-approximate Nash Equilibrium. Quiet surprisingly the approximate NE found by our sub-optimal algorithm slightly Pareto-dominates the exact NE of the game. This observation shows that the performance of proposed algorithm is acceptable while it is relatively easy to implement. Other techniques such as pricing might be useful to improve the efficiency of the overall system. The situation where each user is allowed to freely choose its covariance matrix merely constrained to the maximum power is the natural extension of this paper. Moreover, the discussion over the effect of successive interference cancellation and multiple carrier seems to be complicated and serve as the challenge of the future works.

References

1. Zappone, A., Björnson, E., Sanguinetti, L., Jorswieck, E.: Globally optimal energy-efficient power control and receiver design in wireless networks. IEEE Trans. Sig. Process. **65**, 2844–2859 (2017)
2. Zappone, A., Jorswieck, E.: Energy efficiency in wireless networks via fractional programming theory. Found. Trends® Commun. Inf. Theory **11**(3–4), 185–396 (2015)
3. Zappone, A., Chong, Z., Jorswieck, E.: Energy-aware competitive power control in relay-assisted interference wireless networks. IEEE Trans. Wirel. Commun. **12**, 1860–1871 (2013)
4. Zhang, C., Lasaulce, S., Agrawal, A., Visoz, R.: Distributed power control with partial channel state information: performance characterization and design. IEEE Trans. Veh. Technol. **68**(9), 8982–8994 (2019)
5. Belmega, E.V., Lasaulce, S.: Energy-efficient precoding for multiple-antenna terminals. IEEE Trans. Sig. Process. **59**(1), 329–340 (2011)
6. Saraydar, C.U., Mandayam, N.B., Goodman, D.J.: Pricing and power control in a multicell wireless data network. IEEE J. Sel. Areas Commun. **19**(10), 1883–1892 (2001)
7. Debreu, G.: A social equilibrium existence theorem. Natl. Acad. Sci. **38**, 886–893 (1952)
8. Dinkelbach, W.: On nonlinear fractional programming. Manag. Sci. **13**(7), 492–498 (1967)
9. Varma, V., Lasaulce, S., Debbah, M., Elayoubi, S.: An energy efficient framework for the analysis of MIMO slow fading channels. IEEE Trans. Sig. Process. **61**(10), 2647–2659 (2013)

10. Raghavendra, P.S., Daneshrad, B.: An energy-efficient water-filling algorithm for OFDM systems. In: 2010 IEEE International Conference on Communications (ICC), pp. 1–5, May 2010
11. Richter, F., Fehske, A.J., Fettweis, G.: Energy efficiency aspects of base station deployment strategies for cellular network. In: IEEE Proceedings of VTC, Fall 2009
12. Ng, D.W.K., Lo, E.S., Schober, R.: Energy-efficient resource allocation in multi-cell OFDMA systems with limited backhaul capacity. IEEE Trans. Wirel. Commun. 11(10), 3618–3631 (2012)
13. He, S., Huang, Y., Jin, S., Yang, L.: Coordinated beamforming for energy efficient transmission in multicell multiuser systems. IEEE Trans. Commun. 61(12), 4961–4971 (2013)
14. Venturino, L., Zappone, A., Risi, C., Buzzi, S.: Energy-efficient scheduling and power allocation in downlink OFDMA networks with base station coordination. IEEE Trans. Wirel. Commun. 14(1), 1–14 (2015)
15. Du, B., Pan, C., Zhang, W., Chen, M.: Distributed energy-efficient power optimization for CoMP systems with max-min fairness. IEEE Commun. Lett. 18(6), 999–1002 (2014)
16. Daskalakis, C., Goldberg, P.W., Papadimitriou, C.H.: The complexity of computing a Nash equilibrium. SIAM J. Comput. 39(3), 195–259 (2009)
17. Nash, J.: Equilibrium points in n-person game. Proc. Natl. Acad. Sci. 36(1), 48–49 (1950)
18. Belmega, E.V., Lasaulce, S., Debbah, M.: Power allocation games for MIMO multiple access channels with coordination. IEEE Trans. Wirel. Commun. 8(6), 3182–3192 (2009)

Combined Beam Alignment and Power Allocation for NOMA-Empowered mmWave Communications

Wissal Attaoui$^{(\boxtimes)}$ and Essaid Sabir$^{(\boxtimes)}$

NEST Research group, LRI laboratory, ENSEM,
Hassan II University of Casablanca, Casablanca, Morocco
{w.attaoui,e.sabir}@ensem.ac.ma

Abstract. Millimeter-wave communications have recently attracted significant interest in future wireless networks regarding its wide bandwidth that achieves high data rates. However, the major challenge lies in the beam alignment problem that may induce a substantial loss in the received power, notably when narrow beams are employed. Therefore, this paper jointly addresses the problem of beam alignment and power allocation in a non-orthogonal multiple access (NOMA) mmWave system. Unlike the conventional orthogonal multiple access (OMA), we study the case where two users are joined for NOMA transmission. Next, to mitigate this combined issue, we propose an optimization formulation owing to maximize the sum rate. We compare two types of antennas: sectorized and lemniscate antenna patterns under NOMA and OMA schemes. Simulation results prove the performance of NOMA-lemniscate based beam alignment and power allocation scheme compared to the conventional OMA scheme.

Keywords: mmWave · NOMA · Beam alignment · Power allocation.

1 Introduction

With the rapid proliferation of mobile data traffic and the high demand for user services, a new frequency spectrum is required. Hence, the millimeter-wave frequency within the range of 30–300 GHz is intended to be an efficient strategy to enhance the system capacity [1]. However, deploying the mmWave frequencies in the 5G network remains a significant challenge. It's unable to support mobile operation [2] since the mobility at mmWave frequencies can lead to fast-changing channel conditions where even tiny fluctuations in the environment can change the channel and impact performance. Furthermore, because of the use of large-scale antenna arrays, reliable channel estimation is considered as a challenging task in mmWave communications [3]. Therefore, beam alignment is adopted since it doesn't require the explicit estimation of the channel coefficients, and it can

© Springer Nature Switzerland AG 2020
O. Habachi et al. (Eds.): UNet 2019, LNCS 12293, pp. 82–95, 2020.
https://doi.org/10.1007/978-3-030-58008-7_7

even solve the deafness problem that appears in the case of discovery neighbor. However, for systems with small beams, a misalignment between transmitter and receiver may occur for a substantial loss in the received power; thereby, the initial access can be considerably postponed regarding the considerable time spent to find suitable initial directions [4]. Accordingly, the beamwidth selection is essential for the beam alignment phase. Therefore, the base station (BS) and user seek all beam pairs from the pre-defined codebook and pick the best one to maximize the beamforming gain, and then the chosen beam pair is used for the data transmission phase.

During the initial detection phase, the BS may select the same beam pair to serve multiple users having the same direction [5]. However, the narrow beam may not be the most suitable alternative as it reaches only one user. In this case, each beam is managed to attend only one devoted user [6], and multiple users are handled by Orthogonal Multiple Access (OMA) such as Space Division Multiple Access (SDMA) or Time Division Multiple Access (TDMA). However, time or space resources may not be fully used when various users choose the same beam. In this regard, the Non-Orthogonal Multiple Access (NOMA) method can be used to simultaneously serve multiple users at the same time and with the same frequency resource [7], which can significantly enhance the spectral performance of mobile communication networks [8,9].

Combining NOMA and mmWave technology for 5G cellular communication yields significant profits in terms of sum rates and outage probabilities [10], since NOMA makes the use of available spectrum more efficiently. Besides, the high directionality feature of mmWave propagation makes the user's channels highly correlated, which facilitated the integration of NOMA in mmWave communications. NOMA can be considered as a potent multiple access technique for mmWave networks due to the following reasons:

- The heavily directional beams applied in mmWave communications drive to correlated channels, which may corrupt the performance of OMA schemes, where the necessity of NOMA.
- The application of NOMA in mmWave communications is capable of enhancing the bandwidth efficiency and supporting massive connectivity.

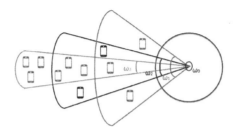

Fig. 1. Beam forming for mmWave NOMA

Compared to conventional OMA, NOMA approves spectrum sharing amongst multiple users, rather than serving a single user in one orthogonal bandwidth block. Hence, mmWave NOMA can serve multiple users in each time slot to support more excellent connectivity and increase the network capacity accordingly [11]. For example, in Fig. 1, the narrow shape beams may serve multiple users at the same time.

There are two main NOMA methods: power-domain and code-domain [9]. Power domain NOMA reaches multiplexing in the power domain, whereas code-domain NOMA accomplishes multiplexing in the code domain. We concentrate on power-domain NOMA, subsequently denoted as NOMA.

This paper addresses the combined beam alignment and power allocation problem in a NOMA mmWave system. We first explain the primary reason for adopting NOMA in 5G; then, we introduce the NOMA scheme during the beam alignment phase [12]. Next, in Sect. 3, we propose the system model recommended for a two-user mmWave NOMA communication, where we study two types of antennas lemniscate and sectorized antennas. In Sect. 4, we formulate the beam alignment problem as a joint beamwidth-power control optimization, where the primary purpose is to maximize the sum rate. Simulation results are presented and analyzed in Sect. 5.

2   Related Works

One of the main issues in mmWave communications is the deafness problem, defined as a misalignment between the transmitter and receiver beams, which requires considerable time in beam search operation [13]. Among recommendations to defeat this problem, authors in [14] proposed a new medium access control (MAC) level collision-notification signal to separate collisions from blockage and deafness, whereas authors in [15] suggested a directional CSMA/CA (Carrier Sense Multiple Access with Collision Avoidance) characterized by spatial reuse protocol in mmWave networks which helps to overcome the deviation issue. The misalignment problem in mmWave networks can be considered as a section of the initial access phase [16], where the mobile user has to discover the base station using a beam alignment procedure. In [17], a beam alignment model is proposed to find out the beam search parameters based on Markov Decision Process; simulation results prove that a bisection search algorithm is better than iterative and exhaustive search algorithms. Whereas, authors in [18] suggest a new search technique called beam switching process, i.e., an initialization process followed by Rosen brock search, aiming to find the optimal beam-pair for data transmission. In [13], authors handle the problems of beamwidth selection and scheduling by proposing two approximation algorithms to maximize the system capacity and the reuse of available spectrum.

3   System Model

We consider a downlink mmWave NOMA communication system in a single cell with one BS and two users (users 1 and 2) as shown in Fig. 2.

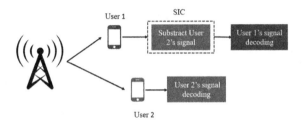

Fig. 2. Downlink NOMA in a single cell with one BS and two users

NOMA uses superposition coding at the transmitter and successive interference cancellation (SIC) at the receiver, therefore multiplexing users in the power domain. The base station (BS) transmits the superposed signals to two users, where user 1 has more significant channel gain than user 2. Under NOMA, the user with bigger channel gain is named strong user, while the user with weaker channel gain noted weak user. The strong user firstly deducts the signal of the weak user within successive interference cancellation (SIC), and then decodes its signal; the weak user identifies its signal immediately and considers the strong user's signal as noise.

We assume that users 1 and 2 have the same angle direction from the BS, then we can consider both users as one receiver. We suggest a frame-slotted system where each frame is divided into two phases, beam alignment phase, and data communication phase. In the first phase, the BS sends beams to users in all directions, and then users transmit him the most suitable pairs of beams through the control channel. Then, the base station forwards the data to users 1 and 2 among the selected beam in the second phase. As shown in Fig. 3, at the origin of each time slot of the detection period, the transmitter T (the base station) sends an initial request containing a tag with beam parameters (P^t, ω^t, G^t), where P^t is the transmission power, ω^t is the beamwidth and G^t is the transmission gain. If the receiver R is located within the beam-shape of ω^t, it detects the beacon signal successfully.

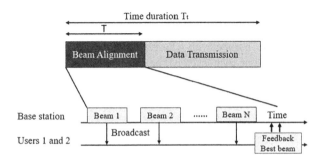

Fig. 3. Beam alignment and data communication phases

In this paper, we address the major challenge of beam alignment in mmWave communication; we focus on the tradeoff between power and beam alignment in NOMA mmWave system. We adopt two types of directional antennas: the sectorized antenna pattern [19] and the lemniscate pattern [20] wherein both, the beamforming gain is a constant for all main lobe angles and a small constant for side lobes. In the following, the new concept dealt with is to consider the directional antenna as a form of an asymmetric lemniscate shape (see Appendix). The transmission and reception gains for the two antenna types are presented in Table 1.

Table 1. Transmission and reception gains

Gain	Antenna type	
	Sectorized antenna [19]	Lemniscate antenna [20]
$G^t(\varphi^t, \omega^t)$	$\begin{cases} \frac{2\pi - (2\pi - \omega^t)\epsilon}{\omega^t} & if\ \varphi^t \le \frac{\omega^t}{2} \\ \epsilon & otherwise \end{cases}$	$\begin{cases} \frac{2}{sin(\omega^t)^2} & if\ \varphi^t \le \frac{\omega^t}{2} \\ \epsilon & otherwise \end{cases}$
$G^r(\varphi^r, \omega^r)$	$\begin{cases} \frac{2\pi - (2\pi - \omega^r)\epsilon}{\omega^r} & if\ \varphi^r \le \frac{\omega^r}{2} \\ \epsilon & otherwise \end{cases}$	$\begin{cases} \frac{2}{sin(\omega^r)^2} & if\ \varphi^r \le \frac{\omega^r}{2} \\ \epsilon & otherwise \end{cases}$

Once the communication between transmitter and receiver is established (the optimal beam pair), next, the BS sends data to users 1 and 2 simultaneously, subject to the constraint of total power. h_1 and h_2 present the channel gains from the BS to users 1 and 2, with $|h_2| < |h_1|$, which indicates that user U_2 holds the weakest instantaneous channel. The NOMA method enables serving users simultaneously by utilizing the whole bandwidth to transmit data with superposition coding technique at the BS and SIC decoding procedure at the users. The BS transmits a linear superposition of the two users' data by allocating a fraction (α_i) of the total power to each user U_i, the power allocated for the i-th user is $P_i = \alpha_i \cdot P$.

The superposed signal transmitted by BS is expressed as:

$$x = \sqrt{P_1} \cdot s_1 + \sqrt{P_2} \cdot s_2 \tag{1}$$

Where s_1 and s_2 are respectively the signals of users 1 and 2.

Thus, γ_1 and γ_2 present respectively the signal to interference plus noise ratio (SINR) of users 1 and 2, as follows:

$$\gamma_1 = \frac{P_1 \cdot h_1 \cdot G^t(\varphi^t, \omega^t) \cdot G^r(\varphi^r, \omega^r)}{\sigma^2}, \tag{2}$$

$$\gamma_2 = \frac{P_2 \cdot h_2 \cdot G^t(\varphi^t, \omega^t) \cdot G^r(\varphi^r, \omega^r)}{P_1 \cdot h_2 \cdot G^t(\varphi^t, \omega^t) \cdot G^r(\varphi^r, \omega^r) + \sigma^2} \tag{3}$$

Where σ^2 is the constant thermal noise power.

Hence, the achievable data rate for users 1 and 2 is given by:

$$R_1 = \left(1 - \frac{T}{T_t}\right) \cdot \log(1 + \gamma_1), \tag{4}$$

$$R_2 = \left(1 - \frac{T}{T_t}\right) \cdot \log(1 + \gamma_2) \tag{5}$$

where:

- T is the total duration of beam searching.
- T_t is the time duration of the two phases (beam alignment and data transmission, see Fig. 2)

4   Joint Beamwidth/Power-Control Optimization

4.1   Beam Alignment

In the Beam alignment phase, to find the maximum rate, the transmitter and receiver have to decide on the optimal refined beams within their sectors. One can find the optimal beam searching based on two features:

- Exhaustive search through searching the optimal combination of transmission and reception directions by a sequence of pilot transmission [13]. Let θ^t, θ^r and ω^t, ω^r be the sector-level and beam-level beamwidths at transmitter and receiver respectively. The total duration time T can be expressed as:

$$T = \left\lceil \frac{\theta^t}{\omega^t} \right\rceil \left\lceil \frac{\theta^r}{\omega^r} \right\rceil \tag{6}$$

  where $\lceil . \rceil$ is the ceiling function.

- Alignment appears when the beam of the transmitter is within the coverage area of the receiver or (and) conversely. The alignment probability π is computed according to the coverage and the aperture of the directional antenna. As the radiation pattern of a directional antenna has a form of an asymmetric lemniscate curve composed by main lobe and side lobe, thus based on Eq. (24) in Appendix, π can be represented as follows:

$$\pi(\omega^t, \omega^r) = \tan\left(\frac{\omega^t}{2}\right)^2 \cdot \tan\left(\frac{\omega^r}{2}\right)^2 \tag{7}$$

4.2   Maximizing Throughput

In this paper, we aim to simultaneously optimize beamwidth and power allocation to maximize the sum rate. In general case, we regard a single-cell downlink scenario where the BS serves N users U_i, with i $\in N = \{1, 2, ..., N\}$. The channels are sorted as $0 \leq |h_1|^2 \leq |h_2|^2 \leq ... \leq |h_i|^2 \leq ... \leq |h_N|^2$, which indicates that user U_i always holds the weakest instantaneous channel. The SINR of user U_i is expressed as:

$$\gamma_i = \frac{P_i \cdot h_i \cdot G^t(\omega^t) \cdot G^r(\omega^r)}{\sum\limits_{k=1}^{i} P_k \cdot h_i \cdot G^t(\omega^t) \cdot G^r(\omega^r) + \sigma^2} \tag{8}$$

The achievable sum rate for each user U_i would be:

$$R_i = \pi(\omega^t, \omega^r) \cdot \left(1 - \frac{T}{T_t}\right) \log(1 + \gamma_i) \tag{9}$$

The optimization problem can be formulated as:

$$\max_{\{\omega^t, \omega^r, P_i\}} \pi(\omega^t, \omega^r) \cdot \sum_{i=1}^{N} \left(1 - \frac{T}{T_t}\right) \log(1 + \gamma_i) \tag{10a}$$

$$s.t. \quad R_i(\omega^t, \omega^r, P_1) \geqslant R_i^{min}, \tag{10b}$$

$$\sum_{i=1}^{N} P_i \leqslant P_{max}, \tag{10c}$$

$$\omega^t_{min} \leqslant \omega^t \leqslant \theta^t, \tag{10d}$$

$$\omega^r_{min} \leqslant \omega^r \leqslant \theta^r, \tag{10e}$$

$$T \leqslant T_t. \tag{10f}$$

Where Eq. (10b) presents the minimum rate required for user U_i, and (10c) is the total transmit power constraint for the BS. Based on these equations, we aim to find the optimal beam pair (ω^t, ω^r) and transmit powers (P_i) to maximize the sum rate.

4.3   Feasible Solutions for Two-User Scenario

In this paper, we study two users for simplicity, but the proposed scheme is also suitable for multiple users. The optimization problem for a two-user scenario can be expressed as:

$$\max_{\{} \omega^t, \omega^r, P_1, P_2\} R_1(\omega^t, \omega^r, P_1) + R_2(\omega^t, \omega^r, P_1, P_2) \tag{11a}$$

$$s.t. \quad R_1(\omega^t, \omega^r, P_1) \geqslant R_1^{min}, \tag{11b}$$

$$R_2(\omega^t, \omega^r, P_1, P_2) \geqslant R_2^{min}, \tag{11c}$$

$$P_1 + P_2 \leqslant P_{max}, \tag{11d}$$

$$\omega^t_{min} \leqslant \omega^t \leqslant \theta^t, \tag{11e}$$

$$\omega^r_{min} \leqslant \omega^r \leqslant \theta^r, \tag{11f}$$

$$T \leqslant T_t. \tag{11g}$$

It is obviously seen that (11) is a non-convex optimization problem which is hard to solve. In the following, we aim to convert it into a convex one.

First, let's compute $G(\omega^t, \omega^r) = G^t(\omega^t) \cdot G^r(\omega^r)$ for both sectorized and lemniscate antennas, we obtain:

$$G(\omega^t, \omega^r) \approx \begin{cases} \frac{2\pi(1-\epsilon)^2}{\omega^t\omega^r} + \epsilon & \text{for sectorized antenna,} \\ \frac{4}{(\omega^t\omega^r)^2} & \text{for lemniscate antenna.} \end{cases} \tag{12}$$

We see that beamwidths in (12) can be formulated as $\omega \triangleq \omega^t\omega^r$. Thus, the beamwidths of the transmitting beamformers or the receiving beamformers (i.e., ω^t or ω^r) are not independently important, but their product (i.e., ω) is the only parameter that determines the optimal solution. Hence, mathematically, we can choose any ω^t (or ω^r) that the system allows, and then set $\omega^r = \frac{\omega^*}{\omega^t}$ (or $\omega^t = \frac{\omega^*}{\omega^r}$), which deliver the same optimal objective. In the same regards, $\pi(\omega^t, \omega^r) \approx \left(\frac{\omega^t\omega^r}{4}\right)^2$.

According to the preliminary analysis, we can reformulate (11) as the following optimization problem:

$$\max_{\{\omega, P_1, P_2\}} R_1(\omega, P_1) + R_2(\omega, P_1, P_2) \tag{13a}$$

$$s.t. \quad R_1(\omega, P_1) \geqslant R_1^{min}, \tag{13b}$$

$$R_2(\omega, P_1, P_2) \geqslant R_2^{min}, \tag{13c}$$

$$P_1 + P_2 \leqslant P_{max}, \tag{13d}$$

$$\omega_{min} \leqslant \omega \leqslant \theta, \tag{13e}$$

$$T \leqslant T_t. \tag{13f}$$

We can adopt a one-dimensional search to find the optimal ω. For a given $\omega = \omega''$, the optimization problem (13) will depend just on the powers P_1 and P_2 as follows:

$$\max_{\{\omega, P_1, P_2\}} R_1(P_1) + R_2(P_1, P_2), \tag{14a}$$

$$s.t. \quad R_1(P_1) \geqslant R_1^{min}, \tag{14b}$$

$$R_2(P_1, P_2) \geqslant R_2^{min}, \tag{14c}$$

$$P_1 + P_2 \leqslant P_{max}, \tag{14d}$$

Based on Eq. (9), P_1 and P_2 can be expressed as:

$$P_1(R_1) = 2^{\frac{R_1}{\pi(\omega'')\left(1 - \frac{T}{T_t}\right)}} \cdot \frac{1}{A_1}, \quad and \tag{15}$$

$$P_2(R_2, P_1) = \left(2^{\frac{R_2}{\pi(\omega'')\left(1 - \frac{T}{T_t}\right)}} - 1\right) \cdot \left(P_1 + \frac{1}{A_2}\right), \tag{16}$$

where $A_1 = \frac{h_1 G(\omega'')}{\sigma^2}$ and $A_2 = \frac{h_2 G(\omega'')}{\sigma^2}$. Replacing (15) in (16), we get:

$$
P_2(R_1, R_2) = 2^{\frac{R_1+R_2}{\Pi\left(1-\frac{T}{T_t}\right)}} \frac{1}{A_1} + \left(2^{\frac{R_2}{\Pi\left(1-\frac{T}{T_t}\right)}} - 1\right)\left(\frac{1}{A_2} - \frac{1}{A_1}\right) \\
- 2^{\frac{R_1}{\Pi\left(1-\frac{T}{T_t}\right)}} \frac{1}{A_1}
\tag{17}
$$

Where $\Pi = \pi(\omega'')$. Based on the above computation, the optimization problem in (13) can be transformed into a convex one depending only on R_1 and R_2 as follows:

$$
\max_{R_1, R_2} \quad R_1 + R_2 \tag{18a}
$$

$$
s.t. \quad R_1 \geqslant R_1^{min}, \quad R_2 \geqslant R_2^{min}, \tag{18b}
$$

$$
\frac{2^{\frac{R_1+R_2}{\Pi\left(1-\frac{T}{T_t}\right)}}}{A_1} + 2^{\frac{R_2}{\Pi\left(1-\frac{T}{T_t}\right)}}\left(\frac{1}{A_2} - \frac{1}{A_1}\right) - \frac{1}{A_2} \leqslant P_{max}, \tag{18c}
$$

We use Lagrange Multiplier to identify a solution of the optimization problem (18). To form the Lagrangian, we introduce multipliers λ_1, λ_2 and ν that correspond respectively to the constraints (18b) and (18c), we have:

$$
\mathcal{L} = R_1 + R_2 + \lambda_1(R_1 - R_1^{min}) + \lambda_2(R_2 - R_2^{min}) \\
+ \nu\left(P_{max} - \frac{2^{\frac{R_1+R_2}{\Pi\left(1-\frac{T}{T_t}\right)}}}{A_1} - 2^{\frac{R_2}{\Pi\left(1-\frac{T}{T_t}\right)}}\left(\frac{1}{A_2} - \frac{1}{A_1}\right) + \frac{1}{A_2}\right). \tag{19}
$$

We extend the method of Lagrange multiplier to the Karush-Kuhn-Tucker (KKT) conditions to find the optimal R_1 and R_2 as follows:

$$
\frac{\partial \mathcal{L}}{\partial R_1} = 1 + \lambda_1 - \nu\frac{2^{\frac{R_1+R_2}{\Pi\left(1-\frac{T}{T_t}\right)}}}{A_1\Pi\left(1-\frac{T}{T_t}\right)} \tag{20}
$$

$$
\frac{\partial \mathcal{L}}{\partial R_2} = 1 + \lambda_2 - \nu\frac{2^{\frac{R_1+R_2}{\Pi\left(1-\frac{T}{T_t}\right)}}}{A_1\Pi\left(1-\frac{T}{T_t}\right)} - \nu\left(\frac{1}{A_1} - \frac{1}{A_2}\right)\frac{2^{\frac{R_2}{\Pi\left(1-\frac{T}{T_t}\right)}}}{\Pi\left(1-\frac{T}{T_t}\right)} \tag{21}
$$

From (20) and (21), we obtain:

$$
R_2 = \Pi\left(1 - \frac{T}{T_t}\right)\log_2\left(\frac{\lambda_2 - \lambda_1}{k\nu}\right) \tag{22}
$$

Where $k = \left(\frac{1}{A_2} - \frac{1}{A_1}\right)\frac{ln(2)}{\Pi\left(1-\frac{T}{T_t}\right)}$.

$$
R_1 = \Pi\left(1 - \frac{T}{T_t}\right)\log_2\left(\frac{(1+\lambda_1)(A_1/A_2 - 1)}{\lambda_2 - \lambda_1}\right) \tag{23}
$$

5   Numerical Results

In this section, simulation results are presented to prove the performance of our proposed system; We compare the new proposed asymmetric lemniscate antenna pattern to the sectorized one under NOMA and OMA mmWave schemes. For comparison, we study an orthogonal multiple access (OMA) scheme, e.g., TDMA, where similar time slots are assigned to Users 1 and 2.

We analyze the impact of beamwidth and power allocation on spectrum efficiency. In Fig. 4, we fix the beamwidth, and we vary the sum rate per power allocation, while in Fig. 5 we fix power allocation and we vary the sum rate per beamwidth ω. The distances between Users 1, 2 and BS are 10 and 30 m, respectively. The noise power σ is set as $-80\,dBm$, and the minimum rate is set as $R_1^{min} = R_2^{min} = 2\,bps/Hz$.

Figure 4 displays the sum rate vs the maximum transmit power of the BS. We set the time duration $T/T_t = 10^{-3}$ and $T/T_t = 10^{-5}$. We see that the sum rate rises with the maximum transmit power where lemniscate antenna with NOMA is better then sectorized antenna. Moreover, one can observe that the longer time duration can bring a higher sum rate due to a greater data transmission time. The NOMA-lemniscate scheme performs better results than the OMA scheme.

Meanwhile, Fig. 5 shows the sum rate versus beamwidth. We set $P_{max} = 25\,dBm$, $\theta_{min} = 0$, it is clearly shown that a greater time duration can achieve higher sum rate. Furthermore, we can also observe that the wider beam gives a higher sum rate for NOMA-lemniscate system and respectively. We notice that

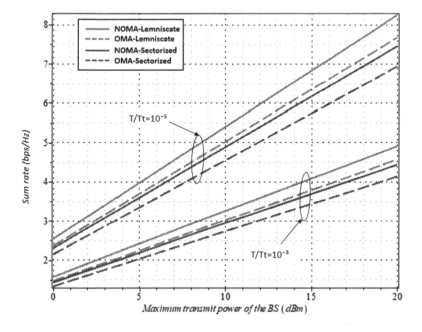

Fig. 4. Sum rate vs maximum transmit power of the BS

the sum-rate following our proposed NOMA-lemniscate system is more important than those under the NOMA-sectorized and conventional OMA schemes.

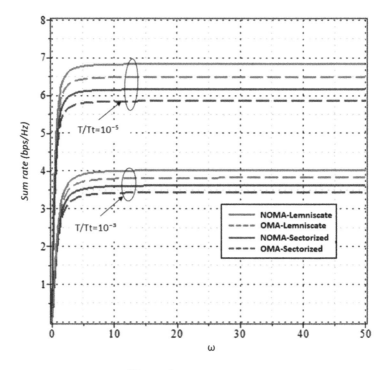

Fig. 5. Sum rate versus ω

6 Conclusion

In this paper, we have studied the beam alignment and power allocation optimization problem for NOMA mmWave communication systems. We employ NOMA in the beam alignment phase, where we seek the maximum sum rate based on two features: exhaustive search and alignment probability. Next, we formulate the problem as a sum-rate maximization under power and beamwidth constraints. Hence, we find a feasible solution for a two-user scenario by converting the optimization problem to a convex one. Simulation results proved that the proposed NOMA-lemniscate scheme performs better results than NOMA-sectorized and OMA schemes.

7 APPENDIX

The directional antenna has a form of an asymmetric lemniscate shape. Therefore, to calculate the alignment probability, we need to find the polar equation of the directional antenna pattern. In this context, we apply the same system described in [21].

7.1 Three Bar Linkage System for Polar Equation of Asymmetric Lemniscate Curve

A geometric design illustrates how we can create a group of asymmetric lemniscate-like curves utilizing a circle and a fixed point. This system presents a sequence of three bars connected to each other with the outside two bars also associated to the plane at two fixed points. The corresponding lengths of these bars, the distance between the fixed points, and the position of the marker (along the middle bar) define the appropriate curve. **Theorem:** For a three-bar linkage system with two fixed points A= (n, 0) and B = (m, 0), and three bars AC, BD, CD such that $dist\,(A,B) = m + n = |CD| = l$ and $|AC| = |BD| = r$, then the family of skewed or asymmetric lemniscate-like curves generated by the marker F with $|CF| = m$ and $|FD| = n$ can be represented by a polar equation ρ:

$$\rho = (m - n)\cos(\Phi) + (m + n).\sqrt{\frac{r^2}{l^2} - \sin(\Phi)^2} \qquad (24)$$

Figure 6 illustrates a simulation of asymmetric lemniscate using MAPLE.

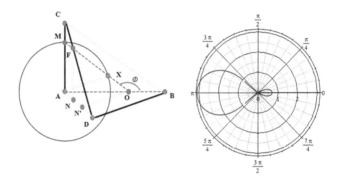

Fig. 6. Geometric Construction and Maple simulation

7.2 Computing Beam Alignment Probability

We calculate the area of the polar equation ρ as:

$$\mathcal{A}(\omega_k) = \int_0^{2\pi} \rho^2 \, d\phi = \pi * \tan(\frac{\omega_k}{2})^2 * (m + n)^2 - 2\pi * mn \qquad (25)$$

Alignment probability transpires when the beam of the transmitter is in the coverage area of the receiver or (and) conversely. In this paper, we will focus on the (and) case. The probability of alignment π depends on beamforming direction, coverage radius and transmission power:

$$\pi\left(\omega^t, \omega^r\right) = \frac{\mathcal{A}\left(\omega^t\right)}{\pi.n^2} \cdot \frac{\mathcal{A}\left(\omega^r\right)}{\pi.n'^2} \qquad (26)$$

Where n is the coverage radius of transmitter and n' is the coverage radius of receiver.

References

1. Cardona, N., Correia, L.M., Calabuig, D.: Key enabling technologies for 5G: millimeter-wave and massive mimo. Int. J. Wireless Inf. Netw. **24**(3), 201–203 (2017)
2. Zhang, S., Wang, G., Chih, I.: Is mmwave ready for cellular deployment? IEEE Access **5**, 14369–14379 (2017)
3. Vlachos, E., Alexandropoulos, G.C., Thompson, J.: Massive mimo channel estimation for millimeter wave systems via matrix completion. IEEE Signal Process. Lett. **25**(11), 1675–1679 (2018)
4. Giordani, M., Mezzavilla, M., Barati, C.N., Rangan, S., Zorzi, M.: Comparative analysis of initial access techniques in 5G mmwave cellular networks, pp. 268–273, March 2016
5. Sun, X., Qi, C., Li, G.Y.: Beam training and allocation for multiuser millimeter wave massive mimo systems. IEEE Trans. Wireless Commun. **18**, 1041–1053 (2019)
6. Xue, Q., Fang, X., Wang, C.: Beamspace SU-MIMO for future millimeter wave wireless communications. IEEE J. Select. Areas Commun. **35**(7), 1564–1575 (2017)
7. Chandra, K., Marcano, A.S., Mumtaz, S., Prasad, R.V., Christiansen, H.L.: Unveiling capacity gains in ultradense networks: using mm-wave noma. IEEE Vehicular Technol. Mag. **13**(2), 75–83 (2018)
8. Xiao, Z., Zhu, L., Gao, Z., Wu, D.O., Xia, X.-G.: User fairness non-orthogonal multiple access (NOMA) for 5G millimeter-wave communications with analog beamforming. CoRR, abs/1811.02908 (2018)
9. Riazul Islam, S.M., Zeng, M., Dobre, O.A.: NOMA in 5G systems: exciting possibilities for enhancing spectral efficiency. CoRR, abs/1706.08215 (2017)
10. Ding, Z., Fan, P., Poor, H.V.: On the coexistence of non-orthogonal multiple access and millimeter-wave communications. In: 2017 IEEE International Conference on Communications (ICC), pp. 1–6, May 2017
11. Xiao, Z., Dai, L., Ding, Z., Choi, J., Xia, P.: Millimeter-wave communication with non-orthogonal multiple access for 5G. CoRR, abs/1709.07980 (2017)
12. Hassan, R.A., Michelusi, N.: Multi-user beam-alignment for millimeter-wave networks. In: 2018 Information Theory and Applications Workshop (ITA), pp. 1–7 (2018)
13. Shokri-Ghadikolaei, H., Gkatzikis, L., Fischione, C.: Beam-searching and transmission scheduling in millimeter wave communications. In: 2015 IEEE International Conference on Communications (ICC), pp. 1292–1297, June 2015
14. Shokri-Ghadikolaei, H., Fischione, C., Popovski, P., Zorzi, M.: Design aspects of short-range millimeter-wave networks: a mac layer perspective. IEEE Network **30**(3), 88–96 (2016)
15. Li, L., et al.: The path to 5G: mmwave aspects. J. Commun. Inf. Netw. **1**(2), 1–18 (2016)
16. Haghighatshoar, S., Caire, G.: The beam alignment problem in mmwave wireless networks. pp. 741–745, November 2016
17. Hussain, M., Michelusi, N.: Throughput optimal beam alignment in millimeter wave networks. In: 2017 Information Theory and Applications Workshop (ITA), pp. 1–6, February 2017
18. Li, B., Zhou, Z., Zou, W., Sun, X., Du, G.: On the efficient beam-forming training for 60 GHZ wireless personal area networks. IEEE Trans. Wireless Commun. **12**(2), 504–515 (2013)

19. Dai, H.-N., Zhao, Q.: On the delay reduction of wireless ad hoc networks with directional antennas. EURASIP J. Wireless Commun. Network. **2015**(1), 16 (2015)
20. Mehdigholi, H., Akbarnejad, S.: Optimization of watt's six-bar linkage to generate straight and parallel leg motion. Int. J. Adv. Robot. Syst. **9**(1), 22 (2012)
21. Porta, G., Nam, H.S., Ju Nam, H.: Characterization of the three-bar linkage system generated symmetric and asymmetric lemniscate-like curves, pp. 56–61, December 2016

Fast Uplink Grant for NOMA: A Federated Learning Based Approach

Oussama Habachi[✉], Mohamed-Ali Adjif[✉], and Jean-Pierre Cances[✉]

XLIM, University of Limoges, Limoges, France
{oussama.habachi,mohamed-ali.adjif,jean-pierre.cances}@xlim.fr

Abstract. Recently, non-orthogonal multiple access (NOMA) technique has emerged and is being considered as a building block of 5G systems and beyond. In this paper, we focus on the resource allocation for NOMA-based systems and we investigate how Machine Type Devices (MTDs) can be arranged into clusters. Specifically, we propose two allocation techniques to enable the integration of massive NOMA-based MTD in the 5G. Firstly, we propose a low-complexity schema where the base station (BS) assigns an MTD to a cluster based on its Channel State Information (CSI) and transmit power in order to ensure that the Successive Interference Cancellation (SIC) can be performed in the uplink as well as the downlink. The proposed technique enable us to allocate an optimal number of MTDs without inter-NOMA-interference (INI), while being of low complexity and communication overhead. In the second framework, we propose a Federated Learning (FL) based-technique using traffic model estimation at the MTD side in order to extend the capacity of the system. In fact, the BS take into account the traffic model of the MTDs in order to use time multiplexing in addition to the power multiplexing to separate MTDs. Then, we propose a synchronization method to allow contending MTDs synchronize their transmissions. Simulation results show that the proposed techniques outperform existing schemes in the literature.

Keywords: Non-orthogonal multiple access · Uplink access

1 Introduction

Telecommunications have experienced a paradigm shift in the past decades with the advent of Internet of things (IoT). With this new huge market, a lot of interesting challenges have emerged, particularly with the potential inclusion of the IoT world in the future fifth generation of cellular mobile communications (5G) and beyond. Indeed, wireless networks have been supporting unprecedented traffic due to the drastic growth of mobile devices, the development of various applications and the implementation of IoT. Consequently, there has been a drastic increase in the number of connected devices. Unlike the third and fourth generation mobile telecommunication systems, where the challenges arose from

© Springer Nature Switzerland AG 2020
O. Habachi et al. (Eds.): UNet 2019, LNCS 12293, pp. 96–109, 2020.
https://doi.org/10.1007/978-3-030-58008-7_8

the demand of high data rate and low latency, the fifth generation (5G) addressed massive connectivity of less sophisticated autonomous wireless devices that may communicate small amounts of data on a relatively infrequent basis. Hence, the explosively increasing demand for wireless traffic cannot be served anymore using orthogonal multiple access (OMA) systems where users share wireless resources in an orthogonal manner. Indeed, the key challenges of the 5G are the higher spectral efficiency, the low latency and the massive connectivity. The latter challenge is particularly hard to address since OMA techniques are suffering from sever congestion problem because of the limited transmission bandwidth.

Specifically, non-orthogonal multiple access (NOMA) techniques have been considered as a promising solutions to tackle the massive demand for bandwidth. In fact, multiple NOMA users are allowed to access the same sub-carrier at the same time using either power domain multiplexing [1,2] or code domain multiplexing [3,4]. Indeed, NOMA requires design of new physical layer and medium access control (MAC) to implement multiple users detection (MUD) technique, such as the successive interference cancellation (SIC), at the receiver side to be able to separate the signals. A plenty of researches have been driven by both academia and industry in order to investigate the design of NOMA technique at the uplink as well as the downlink transmissions. For example, authors of [5–7] proposed an uplink PD- NOMA scheme using random access scheme based on the well-known slotted ALOHA protocol. Specifically, we may consider random access scenario and design multiple access techniques based on contention game and online learning algorithm. For example, [8] proposed a joint resource allocation and power control for random uplink NOMA based on the well-known Multi-Armed Bandit (MAB). After a training period, users are able to determine autonomously the appropriate channel and power level for uplink transmission. On the other hand, uplink NOMA pre-allocation techniques may be considered. For example, Authors of [9] proposed a distributed layered grant-free NOMA framework, in which they divided the cell into different layers based on predetermined inter-layer received power difference to reduce collision probability. In [10], the resource allocation algorithm and user scheduling has been studied for a downlink NOMA where the base station (BS) jointly allocates channels and transmit power. Although several works have investigated the resource allocation for NOMA, allocating an optimal number of users in NOMA networks without inter-NOMA-interference (INI) is still an open challenge that we propose to address in this paper.

Note that taking into account the traffic model of Machine Type Devices (MTDs) while allocating resources enable to design efficient multiple access techniques for NOMA-based WSN. In this paper, we propose a Federated Learning (FL) approach for resources allocation, where MTDs estimate their traffic model and the BS aggregate the parameters to build a global traffic model. In fact, FL is a machine learning attempting to train a centralized model through training distributed low-complexity machine learning over a large number of users, each with unreliable and relatively slow network connections. At every step, local learning algorithm on the users' side are updated, and users communicate the model update to the central server who aggregates data to obtain a new global model. Note that, by using federated learning, the learning task is distributed between the sensors and the BS in order to allow the BS to allocate efficiently Resource blocks (RBs) and power levels. Indeed, with the FL we take advantage of the computation capacity of the BS to aggregate the global machine learning model and from the distributed low-complexity learning algorithms at the sensors side in order to reduce the data exchange and increase the scalability of the system.

The main contributions of the paper are summarized as follows:

- We investigate the joint channel selection and power control problem in Power Domain PD-NOMA and we propose two novel frameworks. Both frameworks take advantage of NOMA to enhance the spectral efficiency.
- Most of the existing NOMA frameworks take the assumption that MTDs are aware of the CSI of other nodes in order to enable the use of the SIC to separate the received signals. In this paper, we release this unrealistic assumption and we propose a new protocol in which the BS informs the MTDs with the CSI of other MTDs in their cluster only. Moreover, we consider that the cluster can be composed of more than two MTDs, unlike what is usually considered in NOMA based techniques.
- We take into account the traffic model of MTDs in order to enhance the capacity of the NOMA-based system. Indeed, we used a FL where the MTDs estimate their traffic model and transmit the parameters to the BS who aggregates the overall traffic model and allocates to each MTD the appropriate resource block and transmit power.

The remainder of the paper is organized as follows. The next section introduces the system model and describes the signals demultiplexing using PD-NOMA. Section 3 proposes a novel resource allocation NOMA-based schema for uplink and downlink transmissions. In Sect. 4, we propose a FL based massive resource allocation schema that take into account the traffic model of the MTDs in order to extend the capacity of the system. Before concluding the paper and giving some perspective, we drive in Sect. 5 an extensive Matlab-based simulation analysis to illustrate the performance of the proposed techniques.

2   System Model

Consider a typical uplink NOMA system, depicted in Fig. 1, composed of M MTDs and a BS. The latter is located at the center of the cell and MTDs are uniformly distributed in the disc with radius r. The MTDs are deployed in the coverage disk of the BS according to a homogeneous PPP Φ_M with density λ_M. Let us focus now on source traffic model for MTDs. We consider that MTDs operate in a regular mode until an event occurs in their environment, where they are triggered into an alarm mode. The event epicenters are represented by a homogeneous PPP Φ_E with density λ_E in the Euclidean plane. The processes Φ_M and Φ_E are assumed independent. We choose to use PPPs because typical nodes can be reasonably assumed to be randomly deployed in the plane, in particular since we are targeting a type of transmission that does not directly involve human intervention.

The available bandwidth is divided into K sub-carriers, and each sub-carrier is divided into W RBs of duration τ. We denote by h_i the channel response from the BS to user i, which is assumed to be zero-mean circular symmetric complex Gaussian random variable with variance σ^2. Since we are using non-orthogonal access, we do not request the M and W to be equal. Indeed, a user can use more than one RB, and the latter will be shared by several users.

Fig. 1. The System model.

Let P_{max} be the maximum transmit power for MTDs, and denote by $p_{i,k}$ the power allocation coefficient of user i on the subcarrier k. The channel between the i-th MTD and the BS on the k-th sub-carrier is denoted by $h_{i,k} = \frac{g_{i,k}}{l_i}$, where $g_{i,k}$ and l_i denotes respectively the Rayleigh fading and the pathloss. The latter is is modelled by Free-Space path loss model [11], i.e. $l_i = \left(\frac{\lambda\sqrt{G_l}}{5\pi d}\right)$, where G_l is the product of the transmit and receive antenna field radiation patterns in the line-of-sight (LOS) direction, and λ is the signal weavelength and d is the distance between MTD and BS. Hence, the received signal on the k-th sub-carrier at the BS is given by:

$$y_k = \sum_{i=1}^{M} h_{i,k}\sqrt{p_{i,k}}s_{i,k} + \sigma \tag{1}$$

where $s_{i,k}$ is the transmit symbol of the MTD i on the sub-carrier k and σ denotes the additive noise at the BS. In order to split the received signal, SIC is carried out at the BS.

Throughout the paper, we assume that each user knows its CSI. In time division duplexing (TDD) mode, the BS may send a beacon signal at the beginning of a time slot to synchronize uplink transmissions. This beacon signal can be used as a pilot signal to allow each user to estimate the CSI.

Consider that user i is multiplexed on the kth sub-carrier, and the transmitted symbol is modulated onto a spreading sequence s_i. Then, the received symbol by BS is expressed as follows:

$$y = \sum_{k=0}^{K}\sum_{i=1}^{M} h_{i,k}\sqrt{p_{i,k}}s_{i,k} + \sigma \tag{2}$$

The BS applies then the SIC in order to separate the superimposed signals. Hence, there is an interesting question that we need to answer: *how to allocate RBs and transmit power to different users in order to make the BS able to separate the signals at the uplink while maximizing the capacity of the system.* The same challenge should be addressed at the downlink as well. In the next section, we propose an allocation technique that addresses the aforementioned challenges.

3   Fast MTD Allocation

In this section, we introduce a low-complexity fast uplink model for MTDs in NOMA-based networks, as illustrated in Fig. 2. First, we consider TDD mode,

---

Algorithm 1. Fast uplink access (FUA)

Initialization: The BS initialize the allocation table CL to $0_{K \times T \times C_{max}}$, where C_{max} is the maximum cluster size
The BS sends a beacon at the beginning of each time slot   **while** *(a new MTD i joins the cell)* **do**

  backoff=0 transmitted=false **while** *(!transmitted)* **do**
    i sends his CSI to the BS using one of the W_c resource blocks **if** *(!transmitted)* **then**
      backoff=backoff+1
      wait(round(random($2^{backoff}$)))

    end

  end
  for $k = 1 : K$ **do**
    $\rho(i,k) = \frac{log(1+((p*HH(k,i))/(sig)))}{max(log(1+((p*HH(1:K,i))/(sig))))}$

  end
  while *!empty($\rho(i,:)$)* **do**
    $k = \arg\max \rho(i,:)$
    for $t = 1 : T$ **do**
      for $p = p_{min} : p_{max}$ **do**
        Find the first j such as $CL(k,t,j) = 0$
        $CL(k,t,j) = \{i, CSI, p\}$
        if $(SIC(CL(k,t,:))! = 0)$ **then**
          Send $CL(\omega,:)$ to i
          exit the algorithm
        else
          $CL(\omega,j) = \{\}$
        end

      end

    end
    $\rho(i,k) = 0$
  end

end
Send NO_ALLOC to user i

---

and we assume that the BS sends a beacon signal at the beginning of a time slot to synchronize transmissions. Hence, the time slot is divided into three part: the beacon, the uplink and the downlink phases. This beacon signal can be used as a pilot signal to allow each MTD to estimate his CSI. Then, we consider that W_c resource blocks are reserved for the contention. In fact, they are used by MTDs when they first join the cell, or when the actual allocation does not meet the MTD's required QoS. The remaining resource blocks are used for transmission. The BS creates a cluster for each resource block, then it allocate MTDs to one or multiple clusters.

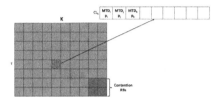

Fig. 2. The proposed resource allocation technique.

The proposed resource allocation schema is depicted as follows:

- **The contention-based access:** When an MTD requests for resource allocation, he should attempt to join the BS through the W_c reserved resource blocks by sending his CSI. Note that the contention-based access is only performed the first time the MTD joins the BS or when he fails to meet its QoS requirements. If the BS fails to decode the MTD's signal, he should retransmit it the next time slot.
- **The resource allocation:** The MTD resource allocation schema is depicted in Algorithm 1. Once the BS receives the signal of the MTD i, it determines his CSI, selects for him his best channel and the lowest power level and checks if he can be allocated to one of the clusters using this channel by executing the SIC. Otherwise, The BS increases the transmit power of the MTDs until reaching P_{max}. Then, the BS selects the second best channel for MTD i and try an allocation. The MTD i is allocated to the first cluster for which the SIC is executed successfully, i.e. the best allocation the BS can afford to him. Then, the MTD allocation is saved in the allocation table of the BS and the corresponding cluster information (CSI and power level of all the MTD in the cluster) are sent back to i. These information are sent to i to enable him performing the SIC at the downlink. An update is sent to all the cluster's members when new MTD joins the cluster. If the BS fails to allocate the MTD to all the clusters, a no-allocation feedback is sent to the MTD i who should wait for a given period before attempting to join the BS again.
- **The uplink phase:** Each MTD who has received an allocation from the BS uses the received transmit power to send his data on the received resource block.

– **The downlink phase:** The BS sends superimposed signals to all the MTDs in the same cluster. They are able to perform SIC since they have received in the initialization phase the CSI and transmit power of all the MTDs in their cluster.

In the next section, we investigate how we can increase the capacity of the NOMA system by taking into account the MTDs' traffic model.

4   Massive MTD Learning-Based Allocation

In this section, we address the massive MTD allocation challenge where an MTD can join a cluster even if the SIC fails. In fact, we take into account the traffic model of MTDs and we use a FL-based approach in order to allow the BS to allocate MTDs.

4.1   Traffic Model

We consider the trafic model, introduced in [12], where the state of an MTD evolve between two states, alarm and regular modes, following a Markov Chain, given in Fig. 3, and the state transition matrix is:

$$P_x = \begin{bmatrix} 1 - \alpha & \alpha \\ 1 - \beta & \beta \end{bmatrix} \tag{3}$$

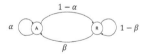

Fig. 3. State transition diagram of the Markov chain model describing the temporal behaviour of the MTD.

This Markov chain is ergodic; it has a unique steady state probability vector $\pi_x = [\pi_x^a; \pi_x^r]$, where π_x^a (π_x^r) is the probability of alarm (regular) state.

We assume that the MTD generates a packet in the alarm (resp. regular) state following a Markov process, illustrated in Fig. 4, and whose the state matrix are given as follows:

$$P_A = \begin{bmatrix} 1 - \alpha_a & \alpha_a \\ 1 - \beta_a & \beta_a \end{bmatrix} \qquad P_R = \begin{bmatrix} 1 - \alpha_r & \alpha_r \\ 1 - \beta_r & \beta_r \end{bmatrix} \tag{4}$$

Hence, the probability that an MTD is active is expressed as follows:

$$\pi_{act} = \frac{\beta_a}{1 + \beta_a + \alpha_a} + \frac{\beta_r}{1 + \beta_r + \alpha_r} \tag{5}$$

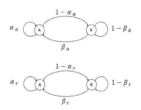

Fig. 4. State transition diagram of the Markov chain model describing the temporal behaviour of the MTD in alarm and regular modes respectively.

4.2   Federated-Learning Algorithm

The proposed algorithm is divided into three step, two of which implemented at the MTD side and one performed by the BS, as illustrated in Algorithm 2. In fact, the traffic model is determined by MTDs who send only their model's parameters to the BS who allocates them both RB and transmit power. Note that learning the traffic model at the MTD side reduces the complexity at the BS side, which increases the scalability of the proposed schema, while having a reasonable complexity to be implemented on low-capacity devices. Moreover, the traffic model learning period is very dependent to the sensed phenomenon, and may be different from an MTD to another one. Thus developing a centralized traffic model learning is very difficult. The joint channel selection and power control is implemented at the BS who has an overall knowledge of the network. Since MTDs are expected to have a relatively long inactivity period, the BS may increase the spectral efficiency by allocating interfering MTDs to the same resources based on their traffic models. Finally, the MTDs use a backoff based algorithm to avoid collisions with other MTDs. The proposed algorithm is explained as follows.

4.3   Traffic Model Learning

We assume that each MTD will monitor his environment in order to learn his traffic model parameters α, β, α_a, β_a, α_r and β_r. These parameters are then transmitted to the BS that will aggregate all the MTDs' traffic model. Note that once the MTD is allocated, he has to ensure that the generated traffic is not higher than the one advertised at the BS. Hence, if an MTD want to increase it's traffic, or the learned traffic model was not accurate, he has to start a new allocation request at the BS with the new traffic model.

4.4   Resource Allocation

Once the BS receives the signal of the MTD i, it determines his CSI, selects for him the lowest power level and try to allocate it to one of the clusters. Indeed, it assumes that the MTD is allocated to this cluster using this power level and executes the SIC. If the SIC fail, the BS determine the set of MTDs in collision with the added one. Then, the BS checks whether the sum of activity probabilities, in Eq. (5), of MTDs in collision is higher than $Prob_{act-max}$. If so, the added MTD cannot be allocated to this cluster. If the BS fails to allocate the MTD to all the cluster, it increases his transmit power level and restart the process. If the allocation is successful, the MTD allocation is saved in the allocation table of the BS and the corresponding resource block and transmit power level are sent back to the MTD. The BS sends also the CSI of all the MTDs who are allocated the same cluster in order to make him able to perform the SIC for the downlink data. An update is sent to all the cluster's members

---

Algorithm 2. Massive MTD allocation

Initialization: The BS initialize the allocation table CL to $0_{|W-W_c|\times C_{max}}$, where C_{max} is the maximum cluster size   The BS sends a beacon at the beginning of each time slot

while *(a new MTD i joins the cell)* **do**
    backoff=0
    transmitted=false
    i observes its environment during a training period (T_{tr} time slots) and estimates his probability of activity π_{act}
    while *(!transmitted)* **do**
        i sends his CSI and π_{act} to the BS using one of the W_c resource blocks
        if *(!transmitted)* **then**
            backoff=backoff+1 wait(round(random($2^{backoff}$)))
        end
    end
    for $p = p_{min} : p_{max}$ **do**
        for $\omega = 1 : |W - W_c|$ **do**
            Find the first j such as $CL(\omega, j) = 0$ and put $CL(\omega, j) = \{i, CSI, p, \pi_{act}\}$
            if $(SIC(CL(\omega, :))! = 0)$ **then**
                Send $CL(\omega, :)$ to i and exit the algorithm
            else
                $T_{coll} = CL(length(CL))$
                for $m = 1 : length(CL)$ **do**
                    if $(!SIC(T_{coll}(1), CL(m))$ **then**
                        $T_{coll} = T_{coll} \cup \{CL(m)\}$
                    end
                end
                if $sum(T_{coll}.\pi_{act})¡Prob_{act-max}$ **then**
                    Send $CL(\omega, :)$ to i
                    exit the algorithm
                else
                    $CL(\omega, j) = \{\}$
                end
            end
        end
    end
end
Send NO_ALLOC to user i

---

also when new MTD joins the cluster. Otherwise, a no-allocation feedback is sent to the MTD who should wait for a long period before attempting to join the BS again.

Algorithm 3. MTDs synchronization

Data: history of observations h,frame_size
Result: The additional delay of MTD i
while *MTD i has packets to transmit* **do**

    i sends the packet and wait for the feedback from the BS

    if *(!transmitted)* **then**

        $T_s = NB_s \cup \{0\}$, //transmission state of node i

        $p_d = \sum\limits_{i=length(T_s)-h}^{length(T_s)} T_s(i)$, //number of successful transmissions during the history period

        $D(i) = \text{mod}(\text{round}(\text{random}(2^{h-p_d})),\text{frame\_size})$

    else

        $T_s = T_s \cup \{1\}$

    end

end

4.5   Traffic Adaptation

Note that the proposed schema increases the capacity of the system, but may result to INI since MTDs transmitting in a given cluster may face collision. Hence, we design a traffic adaptation technique as depicted in Algorithm 2. The idea here is that the BS do not allocate interfering user to the same cluster if the sum of their activity probabilities is higher than $Prob_{act-max}$. Thus , if the SIC fails, the colliding MTDs should arrange theirselves in the frame, by adding some delay, in order to be able to transmit in the same cluster. In fact, if an MTD faces a collision when sending its data, he should add a random delay in order to avoid collision with other MTDs in the same cluster, as illustrated in Fig. 5. Note that the user synchronization is not trivial at all since the SIC outcome is a unique feedback for all the MTDs. Indeed, if MTD i fails the SIC, all weaker MTDs will fail the SIC also even if they are well synchronized, and changing their delay as a reaction to the SIC fail may results in another SIC fail in the future. Hence, we propose that only the first MTD who fails the SIC and MTDs colliding with him will change their delays, other MTDs will ignore the SIC fail. The MTDs' synchronization is depicted in Algorithm 3. Note that the set of colliding MTDs can be easily determined, as we can see in Algorithm 4, since the MTDs have the CSI and transmit powers of their cluster's members.

Algorithm 4. Collision detectiton

Data: Cluster CL
Result: The set of colliding MTDs
$CL_{tmp} = \{\}$
  for $m = 1 : length(CL)$ **do**
    if $(SIC(CL! = 0)$ **then**
    | $U = CL(1)$
    else
    | $CL = CLn\{CL(1)\}$
    end
end
for $m = 2 : length(CL)$ **do**
    if $(!SIC(U, CL(m))$ **then**
    | $CL_{tmp} = CL_{tmp} \cup \{CL(m)\}$
    end
end
return CL_{tmp}

Moreover, in order to increase the stability of the proposed technique, we assume that the more the MTDs transmit successfully, the lower the probability they will change the transmission delay after a collision. Indeed, we consider that the new user who joins the cluster should adapt himself to fit within the available time-slots in the frame. Of course, if he fails during several time slots, all the colliding MTDs will have incentive to change their delay in order to enable all the colliding user to fit into the time slot. Note that the MTDs know that there is a way to fit in the frame since the BS do not allocate interfering MTDs to the same cluster if their activity probabilities is higher than $Prob_{act-max}$, and that MTDs ensure that their generated traffic is coherent with the sent model.

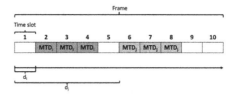

Fig. 5. MTDs i and j user the same resource block and have the same received power, but not at the same time slots. In fact, they use different delays in the frame in order to avoid collision: i is ready to transmit at the beginning of the frame and i is ready to transmit at the third time slot. If there are another MTD having $\pi_{act} < 0.2$, he can be inserted in the cluster, otherwise, i and j should change their starting delays in order to enable the upcoming MTD to transmit with them.

5   Simulation Results

In this section, we drive a Matlab-based simulation in order to evaluate the performance of the proposed resource allocation techniques. As a reference schema, we consider NM-ALOHA in which NOMA is applied in a slotted Aloha basis. This technique was introduced in [7], and has been proved efficient compared to OMA techniques. Hence, in this section we compare the two proposed technique with NM-ALOHA.

We have considered a cell of radius 100 m and $N = 600$ MTDs distributed according tto PPP process of parameter $\lambda_M = 0.01$. Unless specified elsewhere, the average activity period of an MTD is 26%. We have considered that a frame is composed of 30 time-slots.

5.1   System Capacity

In this section, we variate the number of subchannels from 1 to 10 and we variate the number of MTDs from 1 to 600 and we illustrate the probability of user allocation for Fast uplink access (FUA) and for Massive MTD learning-based allocation (MMA). Here, we do not consider the Aloha-based NOMA (NM-ALOHA), since using the latter all MTDs are allowed to transmit randomly.

As we can see in Fig. 6, the system capacity is enhanced by up to 20 times compared to OMA allocation for 200 MTDs and 10 subchannels, and that MMA achieves a better capacity than FUA. Note that the observed capacity enhancement of MMA according to FUA is obtained by taking into account the traffic model of MTDs. Note that with an average activity period of 26%, we may have up to 3 contending MTDs that transmit on the same frame without NOMA interference.

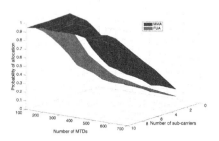

Fig. 6. The allocation probability depending on the number of MTDs and the number of channels in the system.

5.2   Average Throughput

Figure 7 illustrates the average throughout for FUA, MMA and NM-ALOHA. We can observe that even if NM-ALOHA enable more user to transmit, FUA and MMA achieve a far better average throughput per user. This result is somehow expected since NOMA interference is avoided in FUA and MMA.

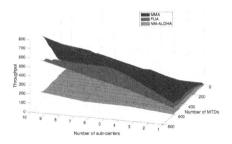

Fig. 7. The average throughput of MTTDs for different N and K.

6   Conclusion

In this paper, we have proposed two novel resource allocation technique in order to jointly allocate channels and transmit power levels in PD-NOMA systems. The first proposed framework allows the BS to allocate the optimal number of user using SIC to separate superimposed signals while ensuring free INI. Moreover, to enable the system to handle more users, we have proposed a novel framework based on a federated learning approach in order to allow the BS and MTDs collaborating to estimate the traffic model and enable massive allocation. We have illustrated, using simulation results, that the learning algorithm converges, and that after a period of adaptation, the system capacity is extended while still being free of INI. Moreover, we have illustrated that taking into account the traffic model enhances significantly the capacity of the system.

References

1. Liu, C., Liang, D.: Heterogeneous networks with power-domain noma: coverage, throughput, and power allocation analysis. IEEE Trans. Wireless Commun. **17**(5), 3524–3539 (2018)
2. Han, W., Ma, X.: Power division multiplexing. In: 2016 IEEE/CIC International Conference on Communications in China (ICCC), pp. 1–6, July 2016
3. Di, B., Song, L., Li, Y., Zhang, S.: Trellis coded modulation for code-domain non-orthogonal multiple access networks. In: 2018 IEEE International Conference on Communications (ICC), pp. 1–6, May 2018
4. Gan, M., Jiao, J., Li, L., Wu, S., Zhang, Q.: Performance analysis of uplink unco-ordinated code-domain NOMA for sins. In: 2018 10th International Conference on Wireless Communications and Signal Processing (WCSP), pp. 1–6, October 2018
5. Ali, M.S., Tabassum, H., Hossain, E.: Dynamic user clustering and power allocation for uplink and downlink non-orthogonal multiple access (NOMA) systems. IEEE Access **4**, 6325–6343 (2016)
6. Zhang, N., Wang, J., Kang, G., Liu, Y.: Uplink nonorthogonal multiple access in 5G systems. IEEE Commun. Lett. **20**(3), 458–461 (2016)
7. Choi, J.: Noma-based random access with multichannel aloha. IEEE J. Select. Areas Commun. **35**(12), 2736–2743 (2017)

8. Choi, J.: Joint channel selection and power control for NOMA: a multi-armed bandit approach. In: MOMENT Workshop, IEEE PIMRC (2019)

9. Jiang, H., Cui, Q., Gu, Y., Qin, X., Zhang, X., Tao, X.: Distributed layered grant-free non-orthogonal multiple access for massive MTC. In: 2018 IEEE 29th Annual International Symposium on Personal, Indoor and Mobile Radio Communications (PIMRC), pp. 1–7, September 2018

10. Di, B., Bayat, S., Song, L., Li, Y.: Radio resource allocation for downlink non-orthogonal multiple access (NOMA) networks using matching theory. In: 2015 IEEE Global Communications Conference (GLOBECOM), pp. 1–6, December 2015

11. Goldsmith, A.: Wireless Commun. Cambridge University Press, New Yor (2005)

12. Thomsen, H., Manchon, C.N., Fleury, B.H.: A traffic model for machine-type communications using spatial point processes. In: 2017 IEEE 28th Annual International Symposium on Personal, Indoor, and Mobile Radio Communications (PIMRC), pp. 1–6, October 2017

Global Modelling of Diffraction Phenomena by Irregular Shapes with Hybrid MOM-GTD Method

Samir Mendil[(✉)] and Taoufik Aguili[(✉)]

Department of Telecommunication Sys'com Laboratory,
National School of Engineering of Tunis Campus, University Farhat Hachad El
Manar of Tunisia, Tunis, Tunisia
myemail.contact@gmail.com, taoufik.aguili@gmail.com

Abstract. In this paper we propose to combine in a hybrid method the moments method (MOM) and the general theory of diffraction (GTD). This hybrid approach is used to analyse any arbitrary shape with multiple and varied dimension also place in free space or in wave guide Some examples, e.g. an antenna mounted near a perfect conductor Complex Object with two plates, demonstrates that the hybrid approach is the most suitable technique for modelling large-scale objects with arbitrary shapes. This approach allows us to resolve the problem, that the other methods can't solve it alone. Generally, random radiation locates on or near an arbitrary form, can be solved using this technique hence the strong advantages of our method.

1 Introduction

Although the technology progress in fast computers and memories capacity are available to communication design's engineers, it remains difficult to analyse some class of electromagnetic scattering problems, such us, bodies with significant dimensions relative to the wavelength, and complex shapes, requiring, accurate analysis of a radiating. during the past several decades, many theoretical models have been constructed to study the scattering mechanisms. for analyse larges forms, the application of numerical method like moment method (MoM), can't be taken into account, because of the memory requirements and the CPU time increase proportionally with the frequency. An application of pure asymptotic techniques such as the Diffraction geometric theory (GTD) remains necessary but not enough. A remedy is found in hybrid methods combining (MOM) method, with asymptotic techniques (GTD) general theory of diffraction. In principle, there is a distinction between asymptotic ray-based techniques, such as GTD techniques, and current-based techniques such as PO. Depending on the structure to be analysed, both have some advantages. a Hybrid MOM/GTD formulations are suitable for problems when we have a structure with large dimension and complex shape e.g. an antenna is located in front of a large scattering body. Our approach is not limited to the bodies of the revolution, but perfectly conducting three-dimensional bodies of arbitrary form can be studied. So, finding an effective method to calculate the electromagnetic scattering

© Springer Nature Switzerland AG 2020
O. Habachi et al. (Eds.): UNet 2019, LNCS 12293, pp. 110–121, 2020.
https://doi.org/10.1007/978-3-030-58008-7_9

by dielectric finite complex object motivated many authors. An exact analytical solution for scattering from a finite object does not exist, several approximations have been proposed [13]. It approximates the induced current in a finite object like cylinder or plate by assuming infinite length. this method is valid for a needle-shaped scattered with a radius much smaller than the wavelength. nevertheless, it should be noted that the solutions of such approximate methods, in general, fail to satisfy the reciprocity theorem.

The purpose of this paper is to demonstrate that the use of the diffraction coefficient highlights the geometric shape, simplify the gain in time and required resources. associated with the numerical method MoM give us an effective method to calculate the electromagnetic scattering by dielectric finite complex object.

2 Hybrid Method

The hybrid technique is used here to solve electromagnetic problems in which antennas or other discontinuities are located on or near a large conductive structure. The basic technique was first introduced in the literature by Thiele and New house [1]. This method involves studying the antenna structure in a moment method format and then modifying the generalized impedance matrix to account for the effects of driving via bodies via GTD.

As a result of the notation used by Thiele and Newhouse, the method of moments is applied to the antenna structure by widening the single current J on the surface a series of basic functions of such: $J_1, J_2, J_3, \ldots$

$$J = \sum_{n=1}^{N} I_n J_n \tag{1}$$

A linear operator L is defined to connect the current distributions to their electric fields. A set of weighting functions W1, W2, W is chosen and an internal product is defined so that:

$$\sum_{n=1}^{N} I_n W_n, L(J_n) = W_m, E^i \tag{2}$$

Where E^i is the incident field on the antenna. This is the same line of the system of N equations described above, by the method of moments. The equation is represented by:

$$[Z](I) = (V) \tag{3}$$

The elements of this impedance matrix are those of the impedance matrix of the free space since only the antenna structure has been considered so far.

These elements are given by:

$$Z_{mn} = \langle W_m, L(J_n) \rangle \tag{4}$$

The scalar product forms a unit space in which;

$$\langle J, aE_1 \rangle = a \langle J, E_1 \rangle + b \langle J, E_2 \rangle \tag{5}$$

Where a and b are complex scalars. If aE_1 in equation represents $L(J_n)$ (which is the field due to J_n and if bE_1 in represents a contribution extra field at Z_{mn} (which is also due to J_n but not because of fields arriving directly)

Then:

$$Z'_{mn} = \langle W_m, aE_1 + bE_2 \rangle \tag{6}$$

$$= \langle W_m, L(J_n) + bL(J_n) \rangle \tag{7}$$

Where a = 1, b = (m, n).

$$Z'_{mn} = \langle W_m, L(J_n) \rangle + \langle W_m, bL(J_n) \rangle = Z_{mn} + Z^g_{mn} \tag{8}$$

The exponent g indicates that one adds to each impedance matrix the term Z^g_{mn} to account for the contributions to the m-th observation point due to J_n scattered fields of the conducting body.

$$[Z'](I') = (V) \tag{9}$$

Where Z' is the generalized impedance matrix correctly modified to take into account the presence of the dispersion body, as well as for the antenna itself. Z^g_{mn} elements are found thanks to the GTD. The solution of the equation is

$$(I') = [Z']^{-1}(V) \tag{10}$$

The antenna is divided into segments no more than one wavelength. These segments are grouped two by two to form modes. A free-space dipole moment method formulation is first performed assuming a piecewise sinusoidal current distribution in a two-segment mode. This test mode current generates a field E which is reacted with all modes of two segments on the dipole. Each of these reactions gives an impedance matrix term.

$$z_{jk} = -\int_{Rec_k} E^i_j(l).I_k(l)dl \tag{11}$$

E^i_j Is the electric field from test mode j to receive mode k.

$I_k(l)$ Is the current expansion distribution also assumed sinusoidal pieces now, we finding the first element of our solution by resolving this integral, it represents our antenna placing only without other objects, after that we propose to calculate the second term Z^g_{mn}.

$$z^g_{jk} = -b_{jk}\int_{Rec_k} E^i_j(l).I_k(l)dl \tag{12}$$

finally, the full solution is:

$$z_{jk} + z^g_{jk} = -\int_{Rec_k} E^i_j(l).I_k(l)dl - b_{jk}\int_{Rec_k} E^i_j(l).I_k(l)dl \tag{13}$$

3   Analyse of Structure's Effect by Hybrid Method

Let considered Fig. 1 resume the problem for an object with curved plates, this modelized a complex object with two rectangular finite plates, Fig. 2 resume the mechanism on the rectangular plates, the two plates have finite length, and the main one has an finite length and radius, in order to highlight the diffraction effect, it's an excellent GTD problem. The complex coefficient b (m, n) is the sum of Reflexion and diffraction coefficient. we will compare our solution with the result of others studies.

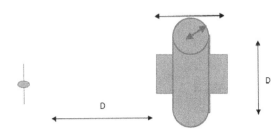

Fig. 1. A general problem.

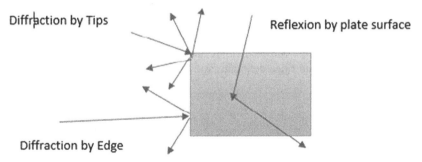

Fig. 2. Diffraction and reflexion mechanism by plate

The antenna is divided into segments no more than one wavelength. These segments are grouped two by two to form. modes. A free-space dipole moment method formulation is first performed assuming a piecewise sinusoidal current distribution in a two-segment mode. This test mode current generates a field E which is reacted with all modes of two segments on the dipole. Each of these reactions gives an impedance matrix term.

By considering the Eq. 13, Now we will applicate the hybrid approach, the phenomena of diffraction/Reflexion by the two plates exposed in Fig. 3 and 4, introduce some coefficients that will be used in method, and we distinguish:

- A diffraction coefficient by a cylinder surface
- A diffraction coefficient by plate edge
- A diffraction coefficient by tips.

The complex coefficient, b in Eq. (5), will be the sum of all these coefficients

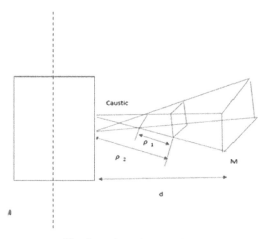

Fig. 3. Asigmatic tube of rays.

For the reflexion Coefficient R define in figure:

$$R = \bar{R} \sqrt{\frac{\rho_1^r \rho_2^r}{(\rho_1^r + d)(\rho_2^r + d)}} \tag{14}$$

d = OM, distance from reflexion point in surface and observation point M. $\bar{R} = \hat{e}_{\parallel}^i \hat{e}_{\parallel}^r - e_\perp e_\perp$ is a dyadic Reflexion coefficient as we shown in Fig. 3.

Diffraction by a curved Edge define in Fig. 4:

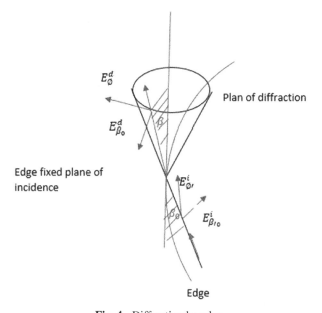

Fig. 4. Diffraction by edge.

We can introduce now the diffraction coefficient for the surface as:

$$D_{s,h}^{surf}(\Phi, \Phi', \beta_0) = \frac{e^{-j\frac{\pi}{4}}}{sin\beta_0}\sqrt{\frac{L}{\pi}}[f(kL, \Phi - \Phi']e^{j2kLcos^2\left(\frac{\Phi-\Phi'}{2}\right)}$$

$$sgn(\pi + \Phi - \Phi') \pm f(kL, \Phi + \Phi']e^{j2kLcos^2\left(\frac{\Phi+\Phi'}{2}\right)}sgn(\pi + \Phi - \Phi')] \qquad (15)$$

With $f(kL, \beta] = \int_{\sqrt{2kL}\left|cos\frac{\beta}{2}\right|}^{\infty} e^{-jz^2}dz$ and $L = s\,sin^2\beta_0$

For the edge the diffraction coefficient is:

$$D_{s,h}^{edge}(\Phi, \Phi', \beta_0) = -\frac{e^{-j\frac{\pi}{4}}}{2\sqrt{2\pi}ksin\beta_0}\left[\frac{F[kL^ia(\Phi - \Phi')]}{cos\left(\frac{\Phi-\Phi'}{2}\right)} \pm \frac{F[F[kL^ra(\Phi + \Phi')]}{cos\left(\frac{\Phi+\Phi'}{2}\right)}\right]$$

$$(16)$$

Angular relationships are expressed by the transition function F (x):
$$2j\left|\sqrt{x}\right|e^{jx}\int_{\left|\sqrt{x}\right|}^{\infty} e^{-jz^2}dz$$

The distance parameter associated with the incident and reflection field is given by

$$L^{i,r} = \frac{s\left(\rho_e^{i,r} + s\right)\rho_1^{i,r}\rho_2^{i,r}sin^2\beta_0}{\rho_e^{i,r}\left(\rho_1^{i,r} + s\right)\left(\rho_2^{i,r} + s\right)}$$

The parameter ρ_e^i is the radius of curvature of the wave front incident at the diffraction point Q_E taken in the plane containing the incident ray and unit vector $\hat{e}$ which is tangent to the edge at Q_E. For the case of spherical waves $\rho_e^i = s'$.

ρ_1^i and ρ_2^i are the principal radii of curvature of the wave front incident to Q_E. Similarly, ρ_1^r and ρ_2^r are the principal radii of the wave front reflected at Q_E, the parameter ρ_e^r is the radius of curvature of the wave front reflected at Q_E taken in the plane containing the ray and e reflected. It is found using:

$$\frac{1}{\rho_e^r} = \frac{1}{\rho_e^i} - \frac{2(\hat{n}.\hat{n}_e)(\hat{I}.\hat{n})}{a_e sin^2\beta_0}$$

The diffraction coefficient for tips given by

$$D_{tips} = \frac{1}{k}\cdot\frac{j}{2.l.n.\frac{2j}{\gamma kbsin\varphi_0}l.n\frac{2j}{\gamma kbsin\varphi}}.l.n\left(\frac{j}{\gamma kbsin\left(\frac{\varphi_0}{2}\right)sin\left(\frac{\varphi}{2}\right)}\right) - \left(\frac{tg\left(\frac{\varphi_0}{2}\right)tg\left(\frac{\varphi}{2}\right)}{cos\varphi + cos\varphi_0}\right)$$

n = number of the plane, b is radius of the finite edge assumed as a cylinder. The other is just defied bellows.also, for two plates:

$$b_{m,n}^1 = 2\left(D^{tips} + D^{edge} + R\right) \qquad (17)$$

and for the central object:

$$b_{m,n}^2 = \left(2D^{edge} + D^{surf}\right) \qquad (18)$$

Finally, the complex coefficient b will be sum for Eq. 17 and 18. In the next section, we will present the numerical application of our approach.

4 Application of Hybrid Method

The system presented before in Fig. 1 is placed we considered the main cylinder C1 of high $R = 2\lambda$, and length $L1 = 7\lambda$, for the two other surfaces of $h2 = \lambda$, $L2 = 2.5\lambda$. the distance of $D = 3\lambda$ from an antenna of length $l = 0.5\lambda$. the frequency f = 3 Ghz for the next subsection.

5 Convergence and Validation of the Method

the radius of antenna a = 0.002 λ and the system with the parameter presented at the begin of this section. We take various steps (N) of discretization N = 75, N = 95, N = 100, N = 120, N = 140. The convergence is obtained at N = 120.

we propose to validate the hybrid method by a reference one, called MoM-eigenfunction [1], The method consists of finding the solution of an axial dipole near an large and infinitely long, perfectly conducting circular cylinder for our case we use 3 plate, the central cylinder placed axially and two others placed perpendicular. A delta impedance matrix representing the effects of the object is found via a moment method procedure. The method incorporates the cylindrical Green's function in the kernel of the integral equation. To carry out this validation we choose two frequencies at low-frequency f = 600 MHz, and at high frequency for f = 3 GHz.

for frequency f = 600 MHz,
Distribution current on the antenna:

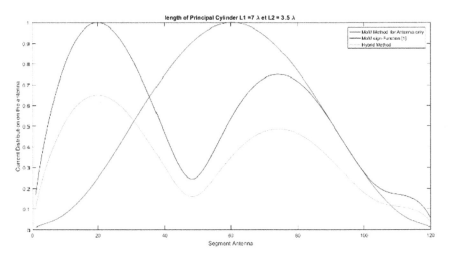

Fig. 5. Current distribution in bas frequency (Color figure online)

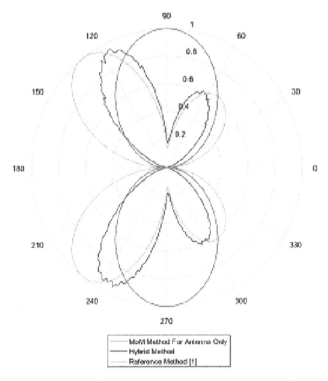

Fig. 6. Pattern diagram in bas frequency (Color figure online)

we notice in Fig. 5 the high of the main and perpendicular plates). A small part of the rays diffracted by the object arrives on the antenna, and the other, the larger one disperses in space. this is very and 6 a difference between the two curves, the hybrid method (green line) and the other, the reference method (blue line), this difference is due to the fact that at low frequency, the value of the wavelength is not enough small relative to the parameters of the object (total length is in order of 7 λ and total width is 3 λ including visible when one sees the distribution of the current on the antenna, this is also explained because the greater intensity of the rays merges their influences are remarkable on the upper part of the antenna. the geometric shape of the object allows that (Fig. 6). the rays that arrive on the upper part of the two secondary plates will be redirected either by diffraction on the antenna and the rays arriving on the lower part of the secondary plates dispersed in space.

for high frequency f = 3 Ghz:
Distribution current on the antenna:

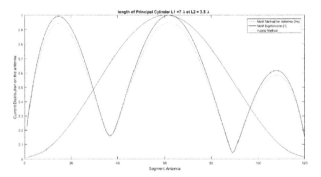

Fig. 7. Current Distribution in high frequency

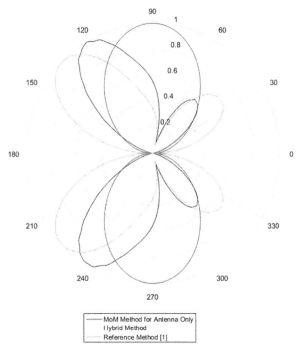

Fig. 8. Pattern distribution in high frequency

Figure 7, 8 shows that the gaps observed at low frequency between the Reference and Hybrid method disappear at high frequency, the two curves are superimposed, we also notice the appearance of a small hump in the shape of the current distribution, it has a half maximum of the two others, as well as the antenna becomes a more directive.

In the next subsection, we analyse the effect of object dimension on the antenna.

6   Electric Characteristic of the Antenna

In this section, we propose to determine some technical characteristics of our antenna (current distribution and radiation diagram) by varying the dimension (length) of the structure presented before Fig. 1.

When we vary Length L1 = from 10 to 10000 λ as length of principal cylinder and L2 = 0.5 λ length of the two secondary rectangular plates.

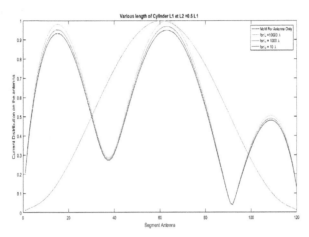

Fig. 9. Current Distribution by changing object length (Color figure online)

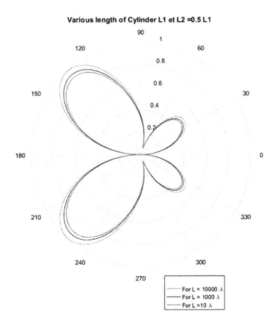

Fig. 10. Pattern distribution (Color figure online)

According to Fig. 9 and 10, by varying the lengths of the plates, it is noted that the intensity of the currents increases in proportion with the lengths of the plates $L = 10 \lambda$ (line blue) at $L = 10000 \lambda$ (yellow line). This is well explained, since the variation of the variables L1 and L2 act as the phenomenon of reflection. The radiation diagram also proves our result.

7   Conclusion

A based hybrid method has been presented that combines MOM with an asymptotic method and GTD. Even though the approach is one of the most important in terms of the shape and size of the scattered, we have the scope of application in this paper to antennas mounted near curved convex surfaces. This particular geometry requires special and complicated features to model if we use another method based on the current flow of the curved convex body and the wire antenna. The asymptotic Method (GTD) has been used to determine the behaviour of electric fields. In which we highlight the importance of Diffraction coefficients for curved edge and curved surface, in our case, and the importance of the use of this coefficient in general case. We were given a solution for calculating the electromagnetic scattering by a dielectric finite complex object. Also, we were given a new hybrid approach, that benefice from the advantage of a numerical method (MoM) and asymptotic one (GTD).

The purpose of this paper is achieved and we demonstrate that the use of the diffraction coefficient highlights the geometric shape, simplify the computation, we gain in time and required resources. Associated with the numerical method MoM give us an effective method to calculate the electromagnetic scattering by dielectric finite complex objects.

Acknowledgments. The authors would like to thank ENIT for providing scholarship during the study.

References

1. Ekelman, E.P., Thiele, G.A.: A hybrid technique for combining the moment method treatment of wire antennas with the GTD for curved surfaces. IEEE Trans. Antennas Propag. **28**, 831–839 (1980)
2. Thiele, G.A.: Overview of selected hybrid methods in radiating system analysis. Proc. IEEE **80**, 67–78 (1992)
3. Bouche, D.P., Molinet, F.A., Mittra, R.: Asymptotic and hybrid techniques for electromagnetic scattering. Proc. IEEE **81**, 1658–1684 (1993)
4. Burnside, W.D., Pathak, P.H.: A summary of hybrid solutions involving moment methods and GTD. In: Applications of the Method of Moments to Electromagnetic Fields, SCEEE Press, St. Cloud (1980)
5. Rao, S.M., Wilton, D.R., Glisson, A.W.: Electromagnetic scattering by surfaces of arbitrary shape. IEEE Trans. Antennas Propag. **30**, 409–418 (1982)
6. Babitch, V.M., Kirpitcnikova, N.Y.: The Boundary-Layer Method in Diffraction Problems. Springer, Heidelberg (1979)

7. Sahalos, J.N., Thiele, G.A.: On the application of the GTD–MM technique and its limitations. IEEE Trans. Antennas Propag. **29**, 780–786 (1981)
8. Medgyesi-Mitschang, L.N., Wang, D.-S.: Hybrid methods for the analysis of complex scatterers. Proc. IEEE **77**, 770–779 (1989)
9. Medgyesi-Mitschang, L.N., Putnam, J.M.: Hybrid formulation for arbitrary 3–D bodies. In: 10th Annual Review of Progress in Applied Computational Electromagnetics, ACES Conference, Monterey, vol. II, pp. 267–274, March 1994
10. Lafitte, O., Lebeau, G.: Equations de Maxwell et opérateur d'impédance sur le bord d'un obstacle convexe absorbant. C. R. Acad. Sci. Paris, T. **316**(Série I), 1177–1182 (1993)
11. Kouyoumjian, R.G., Pathak, P.H.: A uniform geometrical theory of diffraction for an edge in a perfectly conducting surface. Proc. IEEE **62**, 148–1441 (1974)
12. Pathak, P.H., Kouyoumjian, R.G.: The Dyadic Diffraction Coefficient for a Perfectly Conducting Wedge. The Ohio State University, June 1970
13. Zhang, Y.J., Li, E.P.: Fast multipole accelerated scattering matrix method for multiple scattering of a large number of cylinders. Progress Electromagnet. Res. PIER **72**, 105–126 (2007)
14. Valagiannopoulos, C.A.: Electromagnetic scattering from two eccentric metamaterial cylinders with frequency-dependent permittivities differing slightly each other. Progress Electromagnet. Res. B **3**, 23–34 (2008)
15. Illahi, A., Afzaal, M., Naqvi, Q.A.: Scattering of dipole field by a perfect electromagnetic conductor cylinder. Progress Electromagnet. Res. Lett. **4**, 43–53 (2008)
16. Svezhentsev, A.Y.: Some far field features of cylindrical microstrip antenna on an electrically small cylinder. Progress Electromagnet. Res. B **7**, 223–244 (2008)
17. Lai, B.N., Wang, H.B.Y., Liang, C.H.: Progress Electromagnet. Res. **109**, 381–389 (2010)
18. Kuryliak, D.B., Nazarchuk, Z.T., Trishchuk, O.B.: Axially-symmetric TM-waves diffraction by a sphere-conical cavity. Progress Electromagnet. Res. B **73**, 1–16 (2017)
19. Kuryliak, D., Lysechko, V.: Acoustic plane wave diffraction from a truncated semi-infinite cone in axial irradiation. J. Sound Vib. **409**, 81–93 (2017)
20. Kuryliak, D.B.: Axially-symmetric field of the electric dipole over the truncated cone. II. Numerical Modeling. Radio Phys. Radio Astron. **5**(3), 284–290 (2000). (in Russian)
21. Kuryliak, D.B., Kobayashi, K., Nazarchuk, Z.T.: Wave diffraction problem from a semi-infinite truncated cone with the closed-end. Progress Electromagnet. Res. C **88**, 251–267 (2018)

Ubiquitous Internet of Things

Multi-hop LoRa Network with Pipelined Transmission Capability

Dinh Loc Mai and Myung Kyun Kim[(✉)]

University of Ulsan, Daehak-Ro 93, Nam-Gu, Ulsan 44610, South Korea
mkkim@ulsan.ac.kr

Abstract. LoRa (Long Range) has been developed as an attractive IoT (Internet of Things) access network technology in terms of long distance and low power consumption. LoRaWAN a network standard based on LoRa, is an infra-structured network with star topology between LoRa motes, LoRa gateways, and a network server. Even though LoRaWAN has a long distance communication capability, it is difficult to apply LoRa technology in a large area with no communication infrastructure. In those environments, an ad-hoc network based on LoRa can cover a large area easily. In this paper, we propose a multi-hop LoRa network protocol, called *pm-LoRa (pipelined multi-hop LoRa)*, with time-slotted and pipelined transmission capability which can be deployed easily on demand. In our protocol, a converge-cast tree topology is constructed at first for uploading and downloading communication between LoRa motes and a sink node. During the tree construction step, a channel and time-slot is assigned to each tree link, and each node transmits its data on the assigned channel and time-slot during the uploading and downloading data transmission cycles. We developed our multi-hop LoRa node prototype using Multi-tech mDot and made an experiment in the university. The experiment result showed that we could construct a multi-hop LoRa network easily on demand and collect sensor data from LoRa motes with high reliability by removing collisions among neighbor nodes.

Keywords: LoRa networks · Multi-hop networks · Pipelined transmission

1 Introduction

IoT is the extension of Internet connectivity into physical end-devices and every object. IoT has evolved from the convergence of wireless technologies, machine learning, micro-electromechanical systems (MEMS), micro-services and the Internet to connect people and end-devices [1].

LoRa [2] is one of the prominent IoT network technologies in term of long transmission range and low power to meet the needs of IoT applications. With using special modulation technique to counter interference, a LoRa receiver can decode signals 19.5 dB below noise floor, which allows very long communication distance. Petajajarvi et al. [3] addressed a range of 15 km to 30 km in a city, where the transmitter was on the roof of car, and in a boat on open water, meanwhile, the receiver was deployed in a 24 m tall

© Springer Nature Switzerland AG 2020
O. Habachi et al. (Eds.): UNet 2019, LNCS 12293, pp. 125–135, 2020.
https://doi.org/10.1007/978-3-030-58008-7_10

tower. Bor et al. [4] showed that LoRa networks could scale quite well if we use dynamic communication parameter selection and/or multiple sinks. LoRa transceivers available today such as SX1272 [5] use sub-GHz ISM band between 137 MHz to 1020 MHz for transmission. In LoRa motes, the reception capability of a packet depends on the distance, the environment, and the configuration parameters set for radio chip such as SF (spreading factor), BW (bandwidth), and TP (transmission power). These elements affect the RSSI (Received Signal Strength Indicator) value, and transmission is successfully received if the received signal power is greater than the sensitivity threshold of the receiver.

LoRaWAN [6] is a star topology network standardized by LoRa Alliance, where gateways relay messages between end-devices and a central core network server. Basically, LoRaWAN is an infra-structured public network but it will take some time for the LoRaWAN to be accessible by the public. There will also some applications that are difficult to be covered by the public LoRaWAN such as the shipbuilding site under construction with no communication infra-structure. Moreover, the research of [7] shows that, in spite of long transmission range, achieving extensive indoor coverage is still very difficult by a single-hop star topology unless the base stations are deployed with enough density and in proper locations. There are some existing researches tried to build multi-hop MAC protocol using LoRa such as LoRaBlink [8] and LoRa linear network [9] with some limitations.

- In the LoRaBlink protocol, every node will be synchronized by listening to the beacon triggered by the sink node in beacon slots. In data slots, a node which has data to transmit selects the next available data slot and transmits, thus collision may happen to lead low reliability. Synchronizing after finishing an epoch by always listening the beacon consumes much energy.
- The multi-hop LoRa network with linear topology [9] can be used to extend the 1-hop distance of LoRa in an environment with no infra-structure, but is not suitable for the environment in which many nodes are located randomly on a wide area and would have long packet latency from a node to the sink.

This paper proposes a new LoRa MAC protocol, called pm-LoRa, for constructing multi-hop ad-hoc networks to be used in monitoring and control applications, where a public LoRaWAN network is not available. The proposed multi-hop LoRa network uses a TDMA-based pipelined transmission for real-time data transmission from the motes to the sink. The multi-hop LoRa network first constructs a converge-cast tree with the sink node as root and assigns a timeslot and channel to each tree link to allow the pipelined transmission along the path from leaf motes to the sink. When constructing a converge-cast tree topology, we have utilized our LoRa signal capture effect to construct the tree effectively. After constructing the tree, each mote transmits its sensor data through the assigned timeslot and channel after receiving data from its child mote and combining its data with the received data. We have developed a prototype of the LoRa motes using Multi-tech mDot development toolkit and analyzed the performance of the proposed multi-hop LoRa networks by the real experiment in the physical world. We have also analyzed the power consumption in the tree construction period of the multi-hop LoRa network. We compared our protocol with the existing protocol such as LoraBlink [8], LoraWAN [6] and the multi-hop LoRa linear network [9].

- In our protocol, sensor nodes can transmit their data to the sink node in real-time using a pipelined way.
- During the tree construction step, our protocol assigns non-conflicting cells to the neighboring node, which allows no collision when transmitting data. Thus, reliability is higher than the LoRaBlink protocol and power consumption is lower.
- Our protocol uses a collision avoidance mechanism to increase a converge-cast tree construction probability in a general wireless sensor network in a large area.

The remaining part of this paper is organized as follows. Related works on LoRaWAN and multi-hop network technologies based on LoRa are described in Sect. 2. Section 3 describes a multi-hop LoRa network protocol called pm-LoRa proposed in this paper. In Sect. 4, we describe the experimental performance evaluation of pm-LoRa in terms of a converge-cast tree construction probability and message reception reliability. Finally, we conclude the paper in Sect. 5.

2   Lora Communication Networks

LoRaWAN [6] is a one hop network with star topology consisting of LoRa motes, LoRa gateways and a network server. LoRaWAN uses LoRa modulation which allows a very long distance communication with a limited data rate. For communication, a LoRa node has four configuration parameters: carrier frequency, bandwidth, spreading factor, and coding rate [8]. LoRa uses sub-GHz ISM bands as a carrier frequency for communication, which are dependent on the region. LoRa uses 125 kHz, 250 kHz, or 500 kHz as a communication bandwidth. Spreading factor (SF) is the rate between the symbol rate and chip rate. A higher SF increases the receive sensitivity and range but also increases the time on air of the packet and energy consumption.

Many researchers have proposed MAC protocols to extend the LoRaWAN networks to multi-hop mesh networks [7–10]. M. Bor et al. [8] have proposed a multi-hop LoRa MAC protocol called LoRaBlink for low latency communication. LoRaBlink uses slotted channel access based on time synchronization among nodes. The sensor data of a LoRa mote is transmitted to the sink by being forwarded by the nodes with lower hop distance from the sink. The LoRaBlink protocol allows message collisions and takes advantage of the capture effect of LoRa, which incurs a little low message reliability of 80% even in a simple network topology [7]. The LoRaBlink protocol is simple but difficult to be used in a network with many nodes due to a large number of message collisions among neighbor nodes. C. H. Liao et al. [7] have investigated the capture effect of concurrent LoRa transmissions and proposed a CT-offset method to extend the coverage LoRaWAN networks using multi-hop communication through the relay nodes. The CT-offset method can increase the message reception probability when there are multiple concurrent transmissions in LoRa networks. However, the CT-offset method does not utilize the multiple channels of LoRa networks efficiently to reduce the message collision among neighbor nodes. In [9], the authors have proposed a multi-hop MAC protocol to select a proper SF between neighboring nodes and a RPL protocol to choose a routing path with minimum ToA (Time on Air). However, their approach takes long time to construct an RPL tree with minimum ToA.

In [9], the authors have proposed a pipelined transmission scheme in linear multi-hop LoRa networks. The linear multi-hop LoRa network uses a TDMA-based transmission and each node is assigned a timeslot for transmission or reception in a linear network construction step. After constructing a linear multi-hop LoRa network, the leaf node transmits its sensor data in a pipelined way to the sink through forwarding of the intermediate nodes as shown in Fig. 1. We extend this linear multi-hop LoRa networks into networks with tree topology. In multi-hop LoRa networks with a tree topology, the nodes having multiple children combines its own data and the data received from its children and transmits the data in a LoRa packet to reduce the energy and ToA.

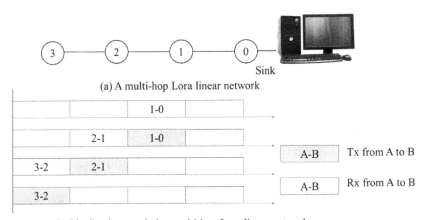

(a) A multi-hop Lora linear network

(b) Pipelined transmission multi-hop Lora linear network

Fig. 1. Pipelined transmission in a linear multi-hop LoRa network.

3   Pipelined Multi-hop Lora Network Protocol with Pipelined Transmission – *Pm-Lora*

The data transmission structure of the proposed multi-hop LoRa networks consists of NCP (Network Construction Period), UTP (Upward Transmission Period) and DTP (Downward Transmission Period) as shown in Fig. 2. In the NCP, each LoRa node joins a multi-hop LoRa network with tree topology and is assigned a channel and time slot to transmit or receive a LoRa packet. After constructing a tree network, data transmission cycles, called major cycles (MCs), are repeated and each MC consists of UTP and DTP. UTP is a period for each node to transmit its data to the sink and consist of n UTCs (Upward Transmission Cycles), where each LoRa mote transmits one data to the sink. After UTP, there is one DTC (Downward Transmission Cycle), for the sink to transmit a command to the LoRa motes.

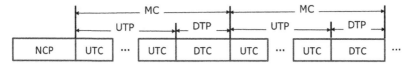

Fig. 2. Data Transmission Structure of Multi-hop LoRa Networks.

3.1 Network Construction Period

In the proposed protocol, sensor data from the motes to the sink is transmitted through the tree constructed during NCP. The tree construction begins from the sink by transmitting INIT message. The tree construction process goes on by exchanging INIT-JOIN-CON-ADV messages among LoRa nodes. A LoRa node sends JOIN message after receiving INIT message from a node already joined. If the joining node receives CON (CONfirm) message from the joined node, it sets the node as a parent and sends ADV (ADVertisement) message by broadcasting to let the neighbor nodes know its channel and timeslot assignment. At first, each node is assigned a unique node ID, and the tree construction process is divided into N cycles, where N is the number of nodes to join. Each cycle consists of 4 timeslots to transmit INIT, JOIN, CON, and ADV messages: T_{S1}, T_{S2}, T_{S3}, and T_{S4}. During constructing tree, each node keeps the following information: P – (parent ID, assigned timeslot and channel); D – depth of the node; Cd – list of (child ID, assigned timeslot and channel); UC – list of cells (timeslot and channel) used by the neighbor nodes. When a node chooses a channel in a timeslot, it selects a channel not conflicting at that timeslot with its neighbors.

After booting, each node try to listen to INIT message and the sink node starts the tree construction process by broadcasting *INIT(D,SID,Cycle,N)* message where D is the depth of the sender, SID is a node ID of the sender, Cycle is the current cycle number, initially 0, N is the number of cycles in one UTC, which is defined by considering the application requirement. The INIT message is transmitted at timeslot S1 of the current cycle. When a LoRa node receives INIT, it sends *JOIN(D,SID,RID,UC)* message at timeslot T_{S2} to join the LoRa network as a child of the sender of INIT message. In JOIN message, SID and RID are node IDs of the sender and receiver, and UC is a cell (channel and timeslot) assignment list of its neighbors. UC is used by the receiver to choose a cell not conflicting with the neighbors. A node collects UC information by overhearing JOIN, CON, and ADV messages during NCP.

If a node receives JOIN message after sending INIT, the node assigns a cell (timeslot Ts and channel Ch) not conflict with the cells in UC list and sends CON message at the next timeslot. The format of *CON(D,SID,RID,Ts,Ch)* message where Ts and Ch is an assigned cell (timeslot, channel). Because every data message from the LoRa motes has to be forwarded to the sink node within one UTC, timeslots are assigned in descending order from the last one. The size of UTC, T_{UTC}, depends on the requirement of the application. We assume $T_{UTC} = N * T_S$, is the size of one data timeslot. If the child node receives CON message from the parent node, it broadcasts *ADV(D,SID,RID,Ts,Ch)* message at the next slot to announce its cell assignment to its neighbor nodes.

If a node sends JOIN message right after receiving INIT, two JOIN messages can collide as shown in Fig. 3-(a). To avoid this kind of collision, each node calculates and waits for a random time within CW (contention window) and sends JOIN message if no node transmits during that time, as explained in the next sub-section. This kind of message collisions can happen between JOIN messages, JOIN and INIT messages, as shown in Fig. 3. If a node loses a contention, the node tries to send JOIN message the next cycle at timeslot T_{S1}.

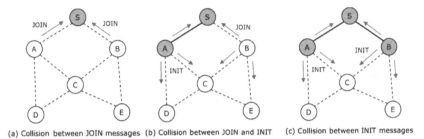

(a) Collision between JOIN messages   (b) Collision between JOIN and INIT   (c) Collision between INIT messages

Fig. 3. Message collisions during NCP.

As an example, we demonstrate the NCP process in Fig. 5 using an example network with 6 nodes in Fig. 4-(a). At first, each node initializes its P, D, Cd, and UC information as shown in Fig. 4-(a). We assume that $T_{UTC} = 5 * T_S$. To start the NCP process, the sink node, S, transmits *INIT(0,S,1,5)* message at slot T_{S1} in the first cycle. After receiving the INIT message, A and B compete to send JOIN message as explained in the next sub-section. Let's assume B wins and B sends *JOIN(1,B,S,{})* message to S to join the network. Node A knows that it lost the competition by overhearing the JOIN message and waits until the next cycle to try to send its JOIN message. After receiving the INIT from B, node S accepts node B as its child and assigns a timeslot TS_5 (the last timeslot in UTC) and a channel C_0 (the first channel) to the link from B to S. Node S transmits *CON(1,S,B,TS_5,C_0)* to B at T_{S3} to inform B its assignment. If node A and B receive the CON message, B sets its parent S and the cell assignment into P and UC information, and A adds (C_0, TS_5) into its UC. In T_{S4}, node B broadcasts *ADV(1,B,TS_5,C_0)* to inform its neighbor nodes the cell assignment. After receiving the ADV message, nodes A, C, and D adds (C_0, TS_5) into their UC list. In T_{S1} of the next cycle, node A and B compete and try to transmit *JOIN(1,A,S,{(C_0, TS_5)})* and *INIT(1,B,2,5)*, respectively. This process is repeated until the network construction completes. Figure 5 shows the NCP process and Fig. 4-(b) shows the information maintained in each node after NCP step.

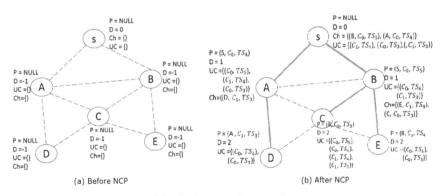

(a) Before NCP   (b) After NCP

Fig. 4. An example network.

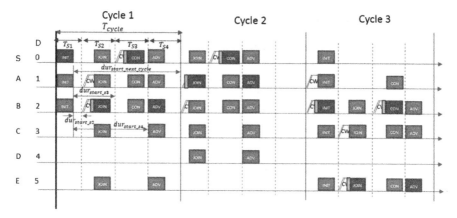

Fig. 5. NCP process in a network of Fig. 4.

3.2 Message Collision Avoidance Mechanism in Pm-LoRa

In pm-LoRa, when two or more nodes transmit INIT or JOIN messages, message collision can occur. In our protocol, we use a collision avoidance mechanism based on CAD (Carrier detection Mechanism) of LoRa transceivers) to increase the probability of network tree construction. In LoRa, the traditional CCA (Clear Channel Assessment) mechanism based on RSSI is not provided by the transceivers because they can decode signals with less power than the noise floor. Instead of that, LoRa transceivers provide a CAD mechanism to detect a preamble signal on the air [5]. The CAD time (T_{CAD}) is about 2 symbol times ($T_{CAD} = 2 * T_S$), where T_S is dependent on the SF as shown in Fig. 6-(b). We use a collision avoidance method based on CAD mechanism of LoRa transceivers. In T_{S1} and T_{S2} timeslots of each cycle in NCP, each node calculates a random delay r ($r = w * T_{CAD}$), where $w = \{0, CW-1\}$, and waits for that time and check the channel before transmitting a message. If a CAD_detected event happens during that time, the node knows it lost the contention, otherwise transmits its message. Figure 6 shows an example when node 1, 2, and 3 contend the channel.

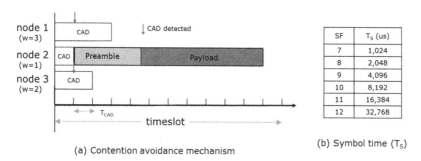

(a) Contention avoidance mechanism

(b) Symbol time (T_S)

Fig. 6. Collision avoidance mechanism in pm-LoRa.

3.3 Data Transmission in Pm-LoRa

In a UTP period of pm-LoRa, each node transmits one data packet in each UTC in a pipelined way to the sink node. If a node receives a packet from its child, it stores the data until all data is received from all of its children, and combines its own data and the data received from its children and transmits the packet using the channel and timeslot assigned during NCP. An sx1272 LoRa transceiver has the maximum Rx buffer with 256B. To allow this kind of pipelined transmission, we restrict the degree (d) and depth (m) of the tree as follows: $d * m \leq \lfloor 256/ \rfloor$, where L is the size of the data generated in each node. Figure 7 shows an upward data transmission of each node in one UTC based on the example of Fig. 5. After n UTCs of pm-LoRa, we have one DTC for the sink node to transmit a command to each node. For downward data transmission, each node including the sink uses the same channel and the time slot in inversed order through the downlink of the tree. Figure 7-(b) shows the downward transmission of the command from the sink in DTC of pm-LoRa.

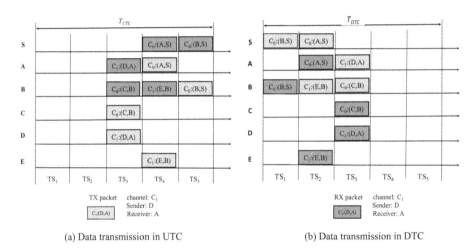

(a) Data transmission in UTC   (b) Data transmission in DTC

Fig. 7. Data transmission example in pm-LoRa.

4   Experimental Evaluation

We have implemented pm-LoRa protocol and developed a pm-LoRa node using Multi-tech mDot module [11] which is consist of Semtech SX1272 LoRa transceiver [5] and a STM32F411RET processor as shown in Fig. 8-(b). To evaluate the performance of pm-LoRa protocol, we made an experiment by deploying 16 nodes with a set of parameters as shown in Table 1.

Table 1. Experiment Parameters.

Number of nodes (M)	16	Bandwidth (BW)	125 kHz
Number of UTCs in UTP	50	Coding rate (CR)	CR4/5
Max. depth	4	Transmission power	0 dBm
Preamble lenth	8 symbols	Header mode	explicit
Spreading factor	SF7	CRC	Enabled

We have measured the performance of successful tree construction of pm-LoRa in terms of contention window size and the successful data reception ratio at the sink after tree construction. The experiment is performed in 2 node deployment scenarios:

- Scenario 1: nodes are randomly placed in the office 5 m × 10 m in the Lab. In this case, every node has a reliable link to each other.
- Scenario 2: nodes are deployed on the university campus area (1000 m × 700 m) as shown in Fig. 8.

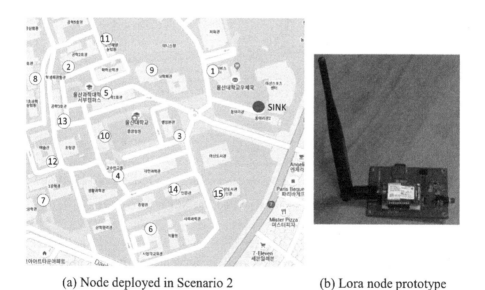

(a) Node deployed in Scenario 2   (b) Lora node prototype

Fig. 8. Node deployment in Scenario 2.

In pm-LoRa protocol, a contention avoidance method using CAD mechanism of LoRa transceivers is used to reduce the message collision probability in the NCP period. The random waiting time in each transmitting node is r = w*T_{CAD}, where w = {0, CW

-1}. We assume the multi-hop LoRa networks are a little sparse and a small CW is can be used. When a small CW is used, the message collision probability can increase. Thus, to increase the successful reception capability of control messages during NCP, we adopt Offset-CT method by adding a short random delay to utilize the capture effect of LoRa [7]. This delay is calculated within 1 symbol time. To measure the effect of contention avoidance mechanism and Offset-CT, we performed an experiment of tree construction in varying CW values and varying number of a maximum number of children.

Figure 9 and Fig. 10 shows the percentages of joined nodes after NCP. As shown in Fig. 9 and Fig. 10, the percentages of joined nodes in NCP increases as CW increase. The result also shows that inserting delay using Offset-CT method increase the percentages of joined nodes.

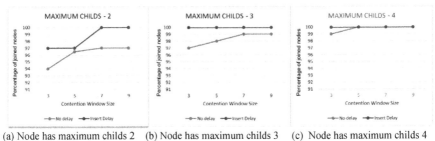

(a) Node has maximum childs 2   (b) Node has maximum childs 3   (c) Node has maximum childs 4

Fig. 9. Percentage of joined nodes during NCP in Scenario 1

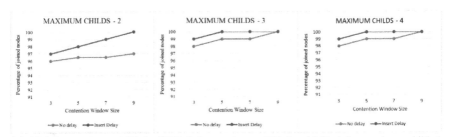

(a) Node has maximum childs 2   (b) Node has maximum childs 3   (c) Node has maximum childs 4

Fig. 10. Percentage of joined nodes during NCP in Scenario 2.

In pm-LoRa protocol, each node is assigned a non-conflicting timeslot and channel for data transmission during NCP. Thus, the data reception probability at the sink is high compared with the traditional aloha-based LoRaWAN protocol. To measure the data reception probability, we have performed a data transmission in 200 cycles in which there is one DTC after 50 UTC. Each node has a 10B sensor data to transmit. After receiving data from all of its children, it attaches its own data and transmits through the assigned timeslot and channel. The results show that data reception probability from all nodes was 97.6% in node deployment Scenario 2 and almost 100% in Scenario 1.

Compared with LoRaBlink [8] protocol, our protocol shows a much higher reliability and low power consumption by eliminating message collision during data transmission.

5 Conclusions

We have proposed a multi-hop LoRa network protocol called pm-LoRa to construct an ad-hoc network in an environment with no LoRaWAN network infra-structure. In pm-LoRa, a converge-cast tree is constructed among LoRa motes and a sink node for upload and download communication. During the tree construction step, a channel and timeslot is assigned to each link and each node transmits its data using the timeslot in a pipelined way. In pm-LoRa, each node can transmits the data reliably and with low latency by removing message collision among neighboring nodes and by using a pipelined transmission from a LoRa mote to the sink node. We have developed a LoRa mote prototype using Multi-tech mDot and evaluated the performance of pm-LoRa protocol in a physical environment. The experimental result shows that pm-LoRa can construct a converge-cast tree successfully and data transmission between LoRa motes and the sink node is reliable. As future work, we are going to analyze the performance of our protocol and improve the performance our protocol in terms of reliability and energy consumption.

Acknowledgement. This research was supported by Basic Science Research Program through the National Research Foundation of Korea (NRF) funded by the Ministry of Education (KNRF-2017030208).

References

1. Ghasempour, A.: Internet of things in smart grid: architecture, applications, services, key technologies, and challenges. Invent. J. **4**(1), 1–12 (2019)
2. LoRa. https://www.semtech.com/lora/what-is-lora. Accessed 02 Mar 2019
3. Petajajarvi, J., Mikhaylov, K., Roivainen, A., Hanninen, T., Pettissalo, M.: On the coverage of LPWANs: range evaluation and channel attenuation model for LoRa technology. In: 2015 14th International Conference on ITS Telecommunications (ITST), pp. 55–59 (2015)
4. Bor, M., Roedig, U., Voigt, T., Aloso, J.: Do LoRa low-power wide-area networks scale. In: Proceedings The 19th ACM International Conference on Modeling, Analysis and Simulation of Wireless and Mobile Systems, pp. 59–67 (2016)
5. Semtech Transceiver SX1272 datasheet. https://www.semtech.com/uploads/documents/sx1272.pdf. Accessed 12 Apr 2018
6. LoRaWAN. https://lora-alliance.org/resource-hub/lorawan-specification-v11. Accessed 04 Dec 2018
7. Liao, C.-H., Zhu, G., Kuwabara, D., Suzuki, M., Morikawa, H.: Multi-hop LoRa networks enabled by concurrent transmission. IEEE Access **5**, 21430–21446 (2017)
8. Bor, M.C., Vidler, J., Roedig, U.: LoRa for the internet of things. EWSN **16**, 361–366 (2016)
9. Duong, C., Kim, M.K.: Reliable multi-hop linear network based on LoRa. Int. J. Control Autom. **11**, 143–154 (2018)
10. Sartori, B., Thielemans, S., Bezunartea, M., Braeken, A., Steenhaut, K.: Enabling RPL multihop communications based on LoRa. In: 2017 IEEE 13th International Conference on Wireless and Mobile Computing, Networking and Communications (WiMob), pp. 1–8 (2017)
11. MultiTech mDot Module. https://www.multitech.com/documents/publications/documents/publications/data-sheets/86002171.pdf. Accessed 12 Apr 2018

IoT-Based Vital Sign Monitoring Using UWB Sensor

Mohamad Mostafa[1], Mohammad Saeed Dayari[1], Somayyeh Chamaani[1(✉)],
Vahid Meghdadi[2], Oussama Habachi[2], and Yannis Pousset[2]

[1] K. N. Toosi University of Technology, 1631714191 Tehran, Iran
{mohamad.mostafa,saeeddayari}@email.kntu.ac.ir,
Chamaani@kntu.ac.ir
[2] Univ. Limoges, CNRS, XLIM, UMR 7252, Limoges, France
meghdadi@ensil.unilim.fr, Oussama.habachi@unilim.fr,
yannis.pousset@univ-poitiers.fr

Abstract. Due to the ageing population and the dramatically increasing of the number of patient or disabled person living alone at home, remote health monitoring has become a critical demand for society, and has attracted the attention of researchers. We implemented an Internet of Things (IoT)–based remote health monitoring prototype using the integration between Xethru Ultra Wide Band (UWB) sensor for data collection and vital sign measurement, Lora protocol for data transmission, Raspberry Pi as gateway for processing, and a server for data storage. In the gateway, the UWB sensor data are processed and cleaned using singular value decomposition (SVD) and Singular Spectrum Analysis (SSA) based algorithm in order to detect and monitor the respiration and heartbeat motion of patient. The IoT based scenario using UWB sensor shows a good ability for older people health monitoring without any obtrusion and privacy violation.

Keywords: Internet of Things · Health monitoring · Raspberry Pi · UWB sensor · Heartbeat rate · Respiration rate

1 Introduction

The world's population is ageing, and according to World Population Prospects the estimated number of older persons will increase from 962 million in 2017 to 2.1 billion in 2050, and more older people are living alone at home. For this reason, it is necessary to provide an efficient solution for the complex health issues as heart attack, chronic diseases, diabetes etc.., and to offer a contiguous health monitoring. Traditional health monitoring techniques are not the best solutions due to high cost, stress and privacy violation caused by conventional scenarios. Thus, the aspect today is to use a remote health monitoring techniques based on wireless sensors that provide the health status of patients as heartbeat rate, breathing activity etc..... The health information can be sent to an assistance system using Internet of Things (IoT). The new approach provides a contiguous health monitoring and the cost supervision by medical personnel can be reduced [1, 2].

© Springer Nature Switzerland AG 2020
O. Habachi et al. (Eds.): UNet 2019, LNCS 12293, pp. 136–145, 2020.
https://doi.org/10.1007/978-3-030-58008-7_11

Internet of things can be defined as an infrastructure of interconnected objects, people, systems and information resources together with intelligent services to allow them to process information of the physical and the virtual world and react [3]. IoT has wide applications such as long-term environmental monitoring [4], home automation and alarm system [5, 6], and health monitoring [1, 6–15].

A review on IoT healthcare monitoring applications was done by Ngueyen et al., and an IoT Tiered architecture (IoTTA) was proposed for transforming sensor data into real time clinical feedback [1]. Saha et al. discussed the IoT based monitoring of heart rate, blood pressure, respiration rate, body temperature, body movement and saline levels using the combination of IoT and Raspberry Pi and the conventional sensors e.g. electrocardiogram (ECG) for heart rate estimation [6, 7]. Manisha et al. experimented IoT-based heart attack detection using wearable built in device or sensor. Not only heart attack was detected, but also other heart diseases has been detected [15]. Kumar et al. monitored patient vital signs and body temperature using Raspberry Pi board. Other works also used Raspberry Pi for IoT-based health monitoring system [10, 13].

To the author's knowledge, all previous works were done by using a conventional wearable sensors such ECG, optical sensors etc. [1, 6–15]. As mentioned above, this obtrusive scenario causes stress for older person and patient and limit their motions and make life more complicated. In this paper, IoT-based health monitoring prototype was implemented using contactless ultra wide band (UWB) sensor for respiration and heart rate estimation, where Arduino and Lora protocol is used for data transmission. After that, the signal processing algorithm is applied to the received sensor data in Raspberry Pi in order to estimate the heart and breathing rate. Finally, the rate of heart and respiration motion versus time is saved at the server.

The remainder of paper is organized as follows. Section 2 presents and discusses IoT-based healthcare monitoring architecture using UWB sensor. Results are presented and discussed in Sect. 3. Concluding remarks are given in Sect. 4

2 Methodology

IoT-based applications can be implemented using different integration and combination of wireless communication systems, devices and protocols for sensing, data transmission and processing. The model architecture used in this paper for IoT-based healthcare monitoring is shown in Fig. 1. The model consists of 5 layers. Each layer is discussed in the next subsections. Figure 2 shows the schematic of IoT based health monitoring prototype.

2.1 Sensing Layer

UWB sensor: X4M03 radar sensor from Novelda [16], shown in Fig. 3, is used in this paper to detect remotely the heartbeat rate and the breathing activity. X4M03 is complete impulse radar system with two bands for transmission, the low band 6–8.5 GHz, and the high band 7.25–10.2 GHz, and maximum frame per second up to 225 frames (sampling frequency on observation time). Figure 3 shows the measurement schematic of vital signs motion tracking by UWB soundings. As shown in Fig. 4 the heart and respiration

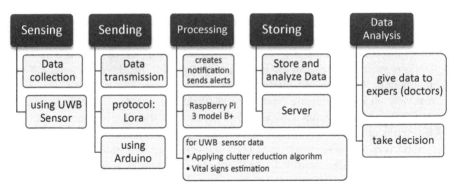

Fig. 1. IoT based healthcare monitoring architecture

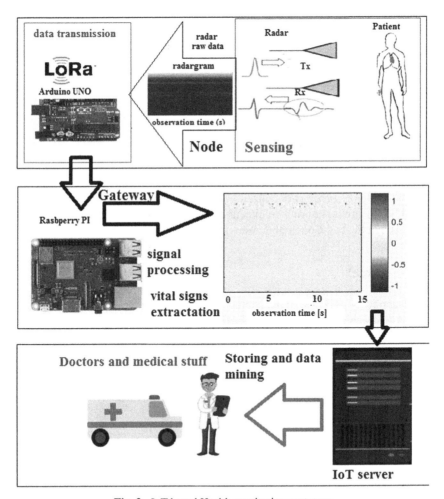

Fig. 2. IoT based Health monitoring prototype

rate can be detected from the reflected signal modulated with the vital signs motion. The reflected or received signal can be written as Eq. (1) [17]:

Fig. 3. X4M03 radar

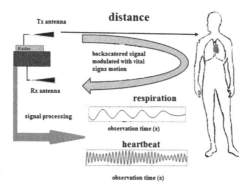

Fig. 4. Measurement schematic of vital sign motion tracking by UWB soundings.

$$y(\tau, t) = x(\tau) * h(t, \tau) = a_v x(\tau - \tau_v(t)) + \sum_i a_i x(\tau - \tau_i) \tag{1}$$

Where $*$ means convolution, $y(\tau, t)$, $x(\tau)$ and $h(t, \tau)$ are the received signal, transmitted signal, and the impulse response respectively, a_v and $\tau_v(t)$ represent the amplitude and time delay of vital signs respectively, a_i and $\tau_i(t)$ represent the amplitude and time delay of other static and non static objects, respectively. The vital signs time delay can be written as:

$$\tau_v(t) = \tau_0 + \tau_r \sin(2\pi f_r t) + \tau_h \sin(2\pi f_h t) \tag{2}$$

$$r(t) = r_0 + A_r \sin(2\pi f_r t) + A_h \sin(2\pi f_h t) \tag{3}$$

Where $r(t)$ models the radar range and r_0 denotes the distance of antenna – chest in case the test person stops breathing, $\tau_0 = 2r_0/v$, $\tau_r = 2A_r/v$ and $\tau_h = 2A_h/v$. $v = 3 \times 10^8$ m/s is the light speed, A_r and f_r represent the amplitude and frequency of respiration motion, where A_h and f_h represent the amplitude and frequency of heartbeat motion [17].

2.2 Sending Layer

Wireless data transmission in network is done by Lora protocol. Lora (Long Range) is a low power, long range and secure platform for IoT applications, which is patented by Semtech. It is a spread spectrum modulation technique derived by chirp spread spectrum (CSS) technology. It uses the sub-gigahertz band and can achieve a sensitivity of over -148 dBm to be capable of sending data in long range (about 8 km) while holding the power consumption very low.

Dragino Lora shield, which is based on SX1276/SX1278, is the chip used in sending layer. For driving dragino shield, Arduino uno is used which is fully compatible with Dragino Lora shield.

The combination of sensor, Arduino and Dragino Lora shield, makes our nodes in network. The gateway, which acts as our processing layer, is a Raspberry Pi with a dragino lora hat, which will be discussed in following.

Sensor node gets the raw data from UWB sensor and due to low power requirement for nodes, it applies no process on them and send them to the gateway for processing.

2.3 Processing Layer

At the processing layer, Raspberry Pi 3 B+ is used. Wireless communication in gateway with nodes, as mentioned before, is done by dragino lora hat. Raspberry Pi is a cheap and capable processor that is used widely and especially on IoT applications. The signal-processing algorithm shown in Fig. 5 is installed on the Raspberry Pi processor; the algorithm is applied to the UWB sensor data in order to remove the clutter and noise from the signal and to estimate the heart and respiration rate. A 6th order Infinite Impulse Response (IIR) high pass Butterworth filter (HPF) with the cutoff frequency of 0.1 Hz is applied in slow time and used to restrict the analysis to the respiration and heartbeat frequency band. After that, simple moving average filter is used to remove the contribution of stationary clutter in the slow time direction, then SVD is applied iteratively to remove the effects of the non stationary clutter and other unwanted signals by removing all the singular values except the 2nd, 3rd, 4th and 5th singular values and reconstructing the matrix after the removal. The bin, r, with the maximum variance is selected as target. Finally, Fast Fourier Transform (FFT) is applied to the range bin r and the heartbeat and respiration rate is obtained. Then, the range bin that contains the vital sign motions can be determined by analyzing the variance of processed data. After that, one dimensional adaptive SSA has been applied to this range. Finally, the respiration and heartbeat rate are estimated by applying Fast Fourier Transform (FFT) to the Reconstructed Components. Using this method, the vital signs motions can be detected and separated from each other and form their harmonics and clutter For more details about the SVD based algorithm, refer to [18]. After processing the raw data in gateway, we send the results to a server to store and analysis them. In addition, a user interface is needed to present data to users.

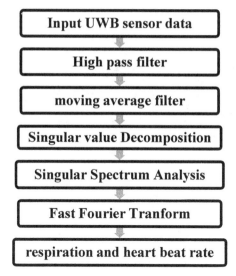

Fig. 5. Block diagram of signal processing algorithm

2.4 Storing Layer

After processes in gateway, we need to store data in a server to be able of monitor and analysis the individual's status. As a free platform we used https://thinger.io to store and show data to user. A token is generated by server to make an access for users and gateway to server.

2.5 Data Analysis

Finally, the medical personnel can have access to the stored data that concerns the health status of patients or older persons, whether the patient is at home or in hospital or any place. This continuous monitoring helps care-giver or physician to take the convenient decision regarding the patients based on the reported data.

3 Experiments and Results

The experiments are done using the combination of: X4M03 Xethru UWB radar for sensing, Arduino using Lora protocol for data transmission, Raspberry Pi 3 B+ for processing and server for data storage. As the sampling frequency equals to 225 Hz in slow time (observation time), time difference between scans is 0.00445 s and the number of scans is equal to 13500 per minute. The length of each scan is 180 range bins. The algorithm mentioned in the previous section is used to remove the unwanted signal such as noise, stationary and non-stationary clutter and applied to the data at the processing layer. The experimental setup and the scenarios for experiments are shown in the Fig. 6.

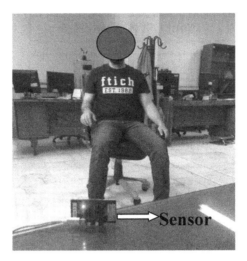

Fig. 6. Experimental setup

The collected radar data before and after applying signal processing algorithm (Fig. 5) is presented in Fig. 7. As shown in Fig. 7, the dominant component before any processing techniques is the clutter and noise, where by applying the SVD based algorithm, the clutter (stationary and non-stationary) and noise are removed and the resulted data becomes cleaner and the respiration and heartbeat rate can be estimated by applying FFT to the SSA reconstructed components.

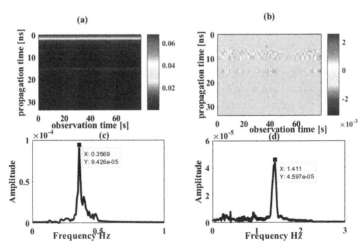

Fig. 7. Collected data (a) before signal processing, (b) after applying signal processing (till SVD). SSA reconstructed components (c) respiration rate, (d) heartbeat rate.

Figure 8 shows the results of measurement scenarios for a person. The measurements are done with different posture for the person under test to simulate the reality. The heartbeat and respiration rate of the different scenarios are shown in Fig. 8. These results show the performance of this IoT-based model to monitor remotely the vital signs of older persons and patients. Since the subject is almost constant, the vital sign variations is very low in a short period of time (15 min).

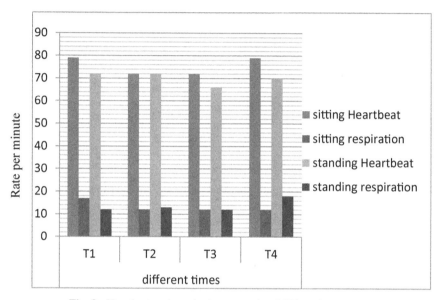

Fig. 8. Heartbeat and respiration rate using IoT-based prototype.

Figure 9 shows the results of measurement scenarios for a person before, during and after exercising. Experiments under different conditions/scenarios are conducted to shows the sensitivity of the prototype in detecting the variation of vital signs rate. The heartbeat and respiration rate of the different scenarios are presented in Fig. 9. To investigate the accuracy, the heartbeat is measured with mobile phone in parallel with UWB sensor. As can be seen, the results of UWB sensor and mobile phone sensors are very close. In Fig. 9, T1 correspond to period before sport, T2 correspond the exercise period, and T3 correspond to resting period after sport. These results show the performance of this IoT-based model to monitor remotely the health of older persons and patients. This experiment proves the concept for long term monitoring of people in real life scenario.

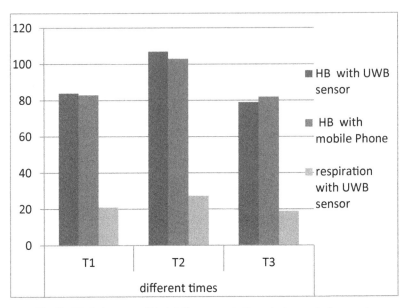

Fig. 9. Vital signs rate (1/min) for person before, during and after sports

4 Conclusion

In this paper, IOT-based remote health monitoring system was proposed and implemented by offering remote sensing for the vital signs via UWB sensor. The sensor data was processed and the noise and clutter has been removed by applying a signal processing algorithm proposed by authors in a previous work; but this time in gateway using Raspberry Pi processor. The vital signs i.e. the heartbeat and the respiration motion were successfully detected and monitored. Finally, the reported data containing the heart and the respiration rate was stored on https://thinger.io server. The proposed model provides monitoring of the health patient condition from the home; reduces the obtrusiveness and probability of human error by using sensors for health data measurement. Furthermore, this technique can detect trends in health status allowing for early identification and intervention before it becomes critical.

References

1. Nguyen, H.H., Mirza, F., Naeem, M.A., Nguyen, M.: A review on IoT healthcare monitoring applications and a vision for transforming sensor data into real-time clinical feedback. In: Proceedings of the 2017 IEEE 21st International Conference on Computer Supported Cooperative Work in Design CSCWD 2017, no. April, pp. 257–262 (2017)
2. Herrmann, R., Sachs, J., Kmec, M., Grimm, M., Rauschenbach, P., IEEE: Ultra-wideband sensor system for remote monitoring of vitality at home. In: 2012 9th European Radar Conference, pp. 234–237 (2012)
3. ISO: Internet of Things (IoT) Preliminary Report 2014 [SWG5]. Education and Training, pp. 1–11 (2015)

4. Lazarescu, M.T.: Design of a WSN platform for long-term environmental monitoring for IoT applications. IEEE J. Emerg. Sel. Top. Circuits Syst. **3**(1), 45–54 (2013)
5. Kodali, R.K., Jain, V., Bose, S., Boppana, L.: IoT based smart security and home automation system. In: Proceeding - IEEE International Conference on Computer Communication Automation, ICCCA 2016, no. October 2017, pp. 1286–1289 (2017)
6. Saha, J., et al.: Advanced IOT based combined remote health monitoring, home automation and alarm system. In: 2018 IEEE 8th Annual Computing and Communication Workshop and Conference CCWC 2018, vol. 2018-January, pp. 602–606 (2018)
7. Saha, H.N., et al.: Health monitoring using Internet of Things (IoT). In: 2017 8th Industrial Automation and Electromechanical Engineering Conference, IEMECON 2017, pp. 69–73 (2017)
8. Kumar, R., Pallikonda Rajasekaran, M.: An IoT based patient monitoring system using raspberry Pi. In: 2016 International Conference on Computing Technologies and Intelligent Data Engineering, ICCTIDE 2016 (2016)
9. Zhang, G., Li, C., Zhang, Y., Xing, C., Yang, J.: SemanMedical: a kind of semantic medical monitoring system model based on the IoT sensors. In: 2012 IEEE 14th International Conference on e-Health Networking, Applications and Services, Healthcom 2012, pp. 238–243 (2012)
10. Pardeshi, V., Sagar, S., Murmurwar, S., Hage, P.: Health monitoring systems using IoT and Raspberry Pi - a review. In: IEEE International Conference on Innovation Industrial Applied Mechanics, ICIMIA 2017, Proceedings, no. ICIMIA, pp. 134–137 (2017)
11. Plageras, A.P., Psannis, K.E., Ishibashi, Y., Kim, B.G.: IoT-based surveillance system for ubiquitous healthcare. In: IECON Proceedings of the Industrial Electronic Conference, pp. 6226–6230, 2016, December 2017
12. Van Os, H.J.A., et al.: Concomitant headache in acute ischaemic stroke: relation with CT angiography and ct perfusion characteristics. Int. J. Stroke **10**, 217 (2015)
13. Koshti, M., Ganorkar, S., M.E. Student: IoT based health monitoring system by using raspberry Pi and ECG signal. Int. J. Innov. Res. Sci. Eng. Technol. (An ISO Certif. Organ.) **3297**(5), 8977–8985 (2016)
14. Manoj, A.S., Hussain, M.A., Teja, P.S.: Patient health monitoring using IoT, pp. 30–45, December 2018
15. Manisha, M., Neeraja, K., Sindhura, V., Ramya, P.: IoT on heart attack detection and heart rate monitoring. Int. J. Innov. Eng. Technol. **7**(2), 459–466 (2016)
16. Novelda. XeThru X4 Radar User Guide. https://www.xethru.com/community/resources/xethru-x4-radar-user-guide.149/
17. Liang, X., Deng, J., Zhang, H., Gulliver, T.A.: Ultra-wideband impulse radar through-wall detection of vital signs. Sci. Rep. **8**(1), 1–21 (2018)
18. Mostafa, M., Chamaani, S., Sachs, J.: Applying singular value decomposition for clutter reduction in heartbeat estimation using M-sequence UWB Radar. In: Proceeding of the International Radar Symposium 2018-June, vol. I, pp. 1–10 (2018)

IIoT-Based Prognostic Health Management Using a Markov Decision Process Approach

Khadija Berhili[1], Mohammed-Amine Koulali[2(✉)], and Yahya Berrehili[1]

[1] Industrial and Seismic Engineering Team, EGIS Laboratory,
National School for Applied Sciences of Oujda,
Mohammed Premier University, Oujda, Morocco
berhilikhadija@gmail.com, yahyaberrehili@yahoo.fr
[2] Modélisation et Simulation Numérique Research Team, M2N Laboratory,
National School for Applied Sciences of Oujda,
Mohammed Premier University, Oujda, Morocco
m.koulali@ump.ac.ma

Abstract. Recent advances in Industrial Internet of Things (IIoT) made them a key component of the Industry 4.0. Thus, several aspects of the latter, such as scheduling maintenance operations, could benefit from the existing IIoT infrastructure. We consider an IIoT-based Prognostic Health Management network for industrial facilities. Our objective is to characterize the optimal maintenance policy that favors grouping maintenance operations while reducing the deterioration and failure costs. We rely on Markov Decision Process with full information Theory to develop a realistic model for the IIoT-based PHM system in an industrial facility with multiple components prone to failure. We investigate the structural properties of optimal policies and provide numerical investigations.

Keywords: Industrial Internet of Things · Prognostic Health Management · Markov Decision Process · Optimal policy · Structural properties

1 Introduction

Prognostics and Health Management (PHM) for maintenance of industrial systems has attracted a vast research effort within the reliability community [1–3]. The deployed endeavors enfold on novel maintenance procedures proposal, degradation processes modeling, health state monitoring/prediction and anticipation of industrial systems failure [4–6].

Increasing reliability and reducing failure-related costs in addition to supplying operators with an integrated vision of their industrial facilities health state are key objectives of PHM. Indeed, PHM frameworks have to trigger alerts to give enough time ahead for failures apparition. Thus, operators are spared the burden of handling the outcomes of failures and can implement coordinated responses either through preventive or corrective maintenance [23].

Industrial maintenance aims at lowering maintenance costs and increasing performance and covers Breakdown Maintenance (BM), Time-based Preventive Maintenance

© Springer Nature Switzerland AG 2020
O. Habachi et al. (Eds.): UNet 2019, LNCS 12293, pp. 146–157, 2020.
https://doi.org/10.1007/978-3-030-58008-7_12

(TPM) and Condition Based Maintenance (CBM). Both TPM and CBM are proactive processes that predict when a device could fail. They can detect early failure signs before it becomes a fact, whereas BM deals with breakdowns in a reactive way [7]. Moreover, TPM establishes a maintenance schedule based on the estimated lifetime of components to anticipate failures and breakdowns.

CBM [8,9] attempts to avoid unnecessary scheduled maintenance operations by scheduling maintenance processes based on the knowledge of the monitored devices health state. Overall, CBM can rise performances and reduce maintenance costs by relying on the industrial components health state evolution.

Industrial Internet of Things (IIoT) in [10–13] is a major pillar of the fourth industrial revolution, it offers innovative solutions to novel industrial applications and services. IIoT fields of application have widened due to huge advances in wireless communications and sensing devices. Nowadays, IIoT has applications in a vast panoply of sectors such as transportation, security and safety, health-care monitoring, robotics, agriculture, smart energy, and industrial facilities automation.

IoT-based CBM [14] constitutes a promising use case of IIoT with major benefits for industrial systems as treated in [14–16]. Indeed, IoT integration into existing CBM frameworks allows monitoring components health state and early detection of associated failure symptoms. The coupling of IIoT and CBM permits automated monitoring and reporting health state evolution. Therefore, updating and planning of maintenance operations become possible.

The authors of [17] considered the problem of minimizing the average inventory level and the average number of back-orders. To achieve those objectives, the solution is based on a Reinforcement Learning (RL) approach and consists of maintaining high service level while carrying as low inventory as possible in a deteriorating manufacturing system.

In [18] the authors studied active and preventive maintenance by proposing a collection of manufacturing production process and off-line prediction using a neural network that recorded outputs lifetime within a specific processing condition.

A concept of device electrocardiogram and deep denoising auto-encoder in [19] are the principal means used to propose a remaining useful life prediction approach of an industrial system. The proposed concept and algorithm combine advanced Artificial Intelligence (AI) and characteristics industrial scenario, in order to reduce dependency on experts' knowledge.

In [20] the authors treated CBM of an aircraft fleet maintenance planning problem as a two-stage dynamic decision-making problem. The division of aircraft into dispatch and standby sets is carried out in order to reduce problem scale. As a solution, they proposed a heuristic hybrid game approach.

In [21] the authors considered a joint decision-making problem concerning prognostics-based replacement decisions and procure the needed service parts. They proposed a one-shot sequential game as a solution approach. It is useful enough in managing cases where the follower's best reply to the leader's strategy.

Kalman filtering is used in [22] to investigate the PHM of aircraft power generators. The authors developed a state estimator and assume constant acceleration to prognosticate the state in specific steps before the due time. Thereby, this information helps to spare sudden failures while reducing the life-cycle cost of the overall system.

The authors of [26] formulated their problem as a Markov Decision Process (MDP [24,25]). They purpose to minimize the packet queuing delay considering the available harvested energy while getting the optimal transmission scheduling policy.

The contribution of the paper is two fold:

- We propose an IIoT-based Prognostics and Health Management framework.
- We provide a mathematical model of the proposed framework and prove the existence of threshold type optimal maintenance policies.

The remaining of this paper is structured as follows: firstly, we describe our system model and its formulation in Sect. 2. Secondly, the formulation of a mathematical model for optimal policy is realized in Sect. 3. Next, Sect. 4 provides a generalized system model with multiple industrial components. Then, Sect. 5 presents simulation results for a single component containing n states. Finally, we conclude the paper.

2 Problem Formulation

We consider an IIoT-based scenario back-boned by a network of N connected devices sharing a wireless medium. Each device monitors an industrial component subject to a failure process with a finite number of health states $n \geq 3$. The device monitors the health state of its associated industrial component through sensors. In Fig. 1, the Markov chain describing the health state evolution of a single component with $n = 3$ is illustrated:

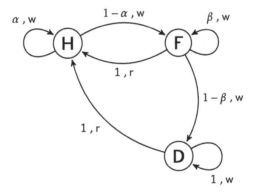

Fig. 1. Markovian model for a single component with three health states ($n = 3$).

Given a particular component with three health states: Healthy (H), Failing (F) and Dead (D), α and β denote the degradation probabilities governing transitions from H to F and F to D respectively. Without loss of generality we assume that $\beta < \alpha$ to account for the effect of accumulated degradation. At most, two actions are available at a given state: wait and replace.

Our objective is to compute an optimal replacement policy for a single monitored component that we will generalize to an industrial plant with multiple components while minimizing maintenance costs through joint scheduling. Moreover, we will investigate the structural properties of the optimal policy to get implementation insights. We seek to elaborate an optimal policy such as when one of the monitored components is in a state of failure, the decision maker will have a policy indicating whether to do immediate individual replacement or to wait for extra time periods and operate joint maintenance on several components. In other words, the optimal policy has to provide a balance between reducing costs through joint maintenance and the risk of incurring a severe failure leading to the dead of components and thus great replacement costs. Time is considered to be discrete and we formulate the optimization problem using a Markov Decision Process.

3   Mathematical Model

In this section, we start with presenting the formal model for a single monitored device as an MDP and analyze the structure of the optimal policy. Next, we generalize our model into N monitored devices.

3.1   Formal Model for a Single Component

To obtain a more fine-grained model for the health state evolution process one could split the two states H and F into sub-states. Our model will be composed of $n > 3$ states as shown in Fig. 2.

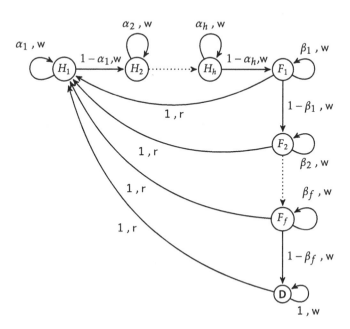

Fig. 2. Markovian model for the component with $n > 3$ Health states.

Finding an optimal policy that optimizes the gain under a Markov cycling environment, while guaranteeing a minimal cost could be formulated as a minimization problem:

$$\min_{\pi} \mathbb{E}[\gamma C_t(s, a)] \tag{1}$$

$C_t(s, a)$ presents the value of one period cost in state s at decision slot t and action a is chosen. Finally, $\mathbb{E}[.]$ is the statistical expectation under transmission policy π. Therefore, the minimization problem (1) for a given component is formulated as a Markov Decision Process, the typical definition of the latter is a five-tuple Γ.

The five-tuple Γ used to model the optimal replacement problem in IIoT-based CBM for industrial facilities.

$$\Gamma \doteq (\mathcal{S}, \mathcal{A}, \mathcal{P}, \mathcal{C}, \gamma) \tag{2}$$

where:

- $\mathcal{S}$ is a finite set of component's health states with cardinality n. Let $\mathcal{H}$ and $\mathcal{F}$ denote respectively the sets of healthy and failing states with cardinalities h and f respectively, and D is the dead state. So $n = h + f + 1$ and:

$$\mathcal{S} = \mathcal{H} \cup \mathcal{F} \cup \{D\} = \{H_1, \ldots, H_h\} \cup \{F_1, \ldots, F_f\} \cup \{D\} \tag{3}$$

- $\mathcal{A}_s$ describes the set of actions available at a given state $s \in \mathcal{S}$:

$$\mathcal{A} = \bigcup_{s \in \mathcal{S}} \mathcal{A}_s = \{w, r\}$$

where action w consists of waiting for the next decision epoch. While, action r stands for replacement which could be either preventive or corrective depending on the actual component degradation state. The following formulas describe the actions available at every component's health state:

$$\forall s \in \mathcal{S} \ \mathcal{A}_s = \begin{cases} w, & s \in \mathcal{H} \\ \{r, w\}, & s \in \mathcal{F} \cup \{D\} \end{cases} \tag{4}$$

- $P : \mathcal{S} \times \mathcal{A} \times \mathcal{S} \longrightarrow [0, 1]$ denotes a stochastic transition matrix governing the component's health state dynamics. Hence, $P(s'|s, a)$ is the probability that a given component's health states evolves from s to s' if action a is undertaken at the current time slot.

$$P(s'|s, w) = \begin{cases} \alpha_i, & \text{if } s = s' \text{ and } s \in \mathcal{H} \\ 1 - \alpha_i, & \text{if } s = H_i \text{ and } s' = H_{i+1}, 1 \le i \le h - 1 \\ 1 - \alpha_h, & \text{if } s = H_h \text{ and } s' = F_1 \\ \beta_i, & \text{if } s = s' \text{ and } s \in \mathcal{F} \\ 1 - \beta_i, & \text{if } s = F_i \text{ and } s' = F_{i+1}, 1 \le i \le f - 1 \\ 1 - \beta_f, & \text{if } s = F_f \text{ and } s' = D \\ 1, & \text{if } s = s' = D \\ 0, & \text{otherwise} \end{cases} \tag{5}$$

$$P(s'|s, r) = \begin{cases} 1, & \text{if } s' = H_1 \text{ and } s \in \mathcal{F} \cup \{D\} \\ 0, & \text{otherwise} \end{cases} \tag{6}$$

Where α_i (respectively $1 - \alpha_i$) is the transition probability from state H_i to it self (respectively H_{i+1}) if action w is chosen for $i \in \{1, \ldots, h\}$. Meanwhile, β_i (respectively $1 - \beta_i$) is the transition probability from state F_i to it self (respectively F_{i+1}) if action w is chosen for $i \in \{1, \ldots, f\}$. $1 - \alpha_h$ (respectively $1 - \beta_f$) denotes the transition probability from state H_h (respectively F_f) to the state F_1 (respectively D).

To take into account the effect of accumulated degradation, we assume that:

$$\begin{cases} \beta_i > \beta_{i+1} \\ \alpha_i > \alpha_{i+1} \\ \alpha_h > \beta_1 \end{cases} \tag{7}$$

- $\mathcal{C} : \mathcal{A} \times \mathcal{S} \longrightarrow \mathbb{R}$ is a real valued mapping representing the running cost function. The cost formula is given by :

$$C(s, w) = \begin{cases} C_o, & \text{if } s \in \mathcal{H} \\ C_o - C_f, & \text{if } s \in \mathcal{F} \\ -C_d, & \text{if } s = D \end{cases} \tag{8}$$

Where C_o, C_f and C_d denote respectively the Operating reward, degradation and failure costs.

$$C(s, r) = \begin{cases} -C_{pr}, s \in \mathcal{F} \\ -C_{cr}, s \in \{D\} \end{cases} \tag{9}$$

Where preventive (respectively corrective) replacement cost is denoted C_{pr} (respectively C_{cr}).
- $\gamma \in]0, 1[$ is the discounting factor.

3.2 Structural Properties of Optimal Policy for a Single Component

Let $\pi^* = (\pi_1^*, \ldots, \pi_n^*)$ denotes the optimal maintenance policy vector. This latter associates to every state a probability distribution on the set of available actions for each health state.

Structured policies are of particular interest due to their nice properties and ease of implementation. We consider a single component with n states and study conditions to establish the existence of optimal threshold maintenance policies.

Theorem 1. *The optimal maintenance problem (2) has an optimal policy of threshold type. Given that the following assumptions hold:*

A.1 We consider the following total order on the space of health states $\mathcal{S} : H_1 < \ldots < H_h < F_1 < \ldots < F_f < D$
A.2 The costs respect the following constraints: $C_f < C_d$ and $C_o + C_{pr} \geq C_{cr}$.

Proof.
- $C(s, a)$ is non-increasing in s for all $a \in \mathcal{A}$. Indeed, $C_a - C_f - C_a = -C_f < 0$ and $C_a - C_d - C_a + C_f = C_f - C_d < 0$ by **A.2**.

– $\forall k \in \mathcal{S}$, $q(k|s,a) = \sum_{j=k}^{n} P(j|s,a)$, is non-decreasing in s for all $k \in \mathcal{S}$ and $a \in \mathcal{A}$.

Indeed, for $(s_{i+2}, s_{i+1}, s_i) \in \mathcal{H}^3$:

$$q(k|s_{i+1}, w) - q(k|s_i, w) = \sum_{j=k}^{D} P(j|s_{i+1}, w) - P(j|s_i, w)$$

$$= P(s_{i+2}|s_{i+1}, w) + P(s_{i+1}|s_{i+1}, w) - P(s_{i+1}|s_i, w) - P(s_i|s_i, w)$$

$$= 0$$

and

$$q(k|s_{i+1}, r) - q(k|s_i, r) = 0 \tag{10}$$

By (10) and (10) we conclude that q is constant for all $k \in \mathcal{S}$ and $a \in \mathcal{A}$ and thus non-decreasing.

– $C(s,a)$ is a super-additive function on $\mathcal{S} \times \mathcal{A}$,

$$\Delta C = C(s_{i+1}, r) + C(s_i, w) - C(s_{i+1}, w) - C(s_i, r)$$

$$= \begin{cases} 0, & (s_{i+1}, s_i) \in \mathcal{H}^2 \\ C_o + C_d - C_f + C_{pr} - C_{cr}, & s_i = F_f \text{ and } s_{i+1} = D \end{cases} \tag{11}$$

Under the condition $C_o + C_{pr} \geq C_{cr}$ we have $\Delta C \geq 0$ and consequently $C(s,a)$ is a super-additive on $\mathcal{S} \times \mathcal{A}$.

– $\sum_{j=H_1}^{D} p(j|s,a)u(j)$ is a super-additive function on $\mathcal{S} \times \mathcal{A}$ for non-increasing u,

$$\Delta Q = \sum_{j=H_1}^{D} p(j|s_{i+1}, r)u(j) + \sum_{j=H_1}^{D} p(j|s_i, w)u(j) - \sum_{j=H_1}^{D} p(j|s_{i+1}, w)u(j) - \sum_{j=H_1}^{D} p(j|s_i, r)u(j)$$

$$= (p(H_1|s_{i+1}, r) - p(H_1|s_i, r))u(H_1) + p(s_{i+1}|s_i, w)u(s_{i+1}) + p(s_i|s_i, w)u(s_i)$$
$$\quad - p(s_{i+2}|s_{i+1}, w)u(s_{i+2}) - p(s_{i+1}|s_{i+1}, w)u(s_{i+1})$$

$$= (p(H_1|s_{i+1}, r) - p(H_1|s_i, r))u(H_1) + (1 - \alpha_i)u(s_{i+1}) + \alpha_i u(s_i)$$
$$\quad - (1 - \alpha_{i+1})u(s_{i+2}) - \alpha_{i+1}u(s_{i+1})$$

$$= \alpha_i(u(s_i) - u(s_{i+1})) + (1 - \alpha_{i+1})(u(s_{i+1}) - u(s_{i+2}))$$

$$\geq 0 \tag{12}$$

Consequently, $\sum_{j=H_1}^{D} p(j|s,a)u(j)$ is a super-additive function on $\mathcal{S} \times \mathcal{A}$ for non-increasing u.

We conclude by virtue of **Theorem 6.11.7** [24] an optimal stationary non-decreasing policy π^* on the set of states $\mathcal{S}$ exists. Since only two actions are available, the optimal maintenance policy is of threshold type.

$$\forall s \in \mathcal{S}, \pi^* = \begin{cases} w, & s < s^* \\ r, & s \geq s^* \end{cases} \tag{13}$$

Where s^* represents the threshold value.   ∎

4   Generalized System Model

In the previously proposed model, we restricted our analysis to a single monitored component. In this section, we will generalize our proposed model to N monitored components with $n \geq 3$ health states. This multi-component maintenance model accounts for costs reduction through grouping of the maintenance periods for several components. The proposed model balances the gain of grouping maintenance operations while coping with the increase of deterioration costs due to maintenance delaying. The optimization problem for multiple components is formulated as a Markov Decision Process:

$$\Gamma_N \doteq (\mathcal{S}_N, \mathcal{A}_N, \mathcal{P}_N, \mathcal{C}_N, \gamma) \tag{14}$$

- $\mathcal{S}_N \doteq \mathcal{S}^N = \displaystyle\underset{i=1}{\overset{N}{\times}} \mathcal{S}$ is the Cartesian product of N health state spaces $\mathcal{S}$.

- $\mathcal{A}_N \doteq \mathcal{A}^N = \displaystyle\underset{i=1}{\overset{N}{\times}} \mathcal{A}$ is the Cartesian product of N action spaces $\mathcal{A}$.

- $\mathcal{P}_N(s'|s,a) = \displaystyle\prod_{i=1}^{N} P(s_i', s_i, a_i)$ represents the states dynamics. It is noteworthy to mention that state deterioration processes are supposed to evolve independently.

- $\mathcal{C}_N(s,a)$ is the reward utility Choosing a joined maintenance strategy for a set of components allows for cost reduction. Indeed, incurred stop times and maintenance operator costs are shared efficiently to deal with multiple failing components simultaneously.

$$\mathcal{C}_N(s,a) = \sum_{i=1}^{N} \mathbb{1}_{\{a_i=w, s_i \in \mathcal{H}\}} C_o + \times (\mathbb{1}_{\{a_i=w, s_i \in \mathcal{F}\}} \times (C_o - C_f) - \mathbb{1}_{\{a_i=w, s_i=D\}} \times C_d)$$
$$- \frac{C_{pr}}{\displaystyle\sum_{i=1}^{N} \mathbb{1}_{\{a_i=r, s_i \in \mathcal{F}\}}} - \frac{C_{cr}}{\displaystyle\sum_{i=1}^{N} \mathbb{1}_{\{a_i=r, s_i=D\}}} \tag{15}$$

Where $\displaystyle\sum_{i=1}^{N} \mathbb{1}_{\{a_i=w\}}$ is a number of states chosen a wait action, $(\mathbb{1}_{\{s_i \in \mathcal{F}\}} \times C_{deg} + \mathbb{1}_{\{s_i=D\}} \times C_f)$ represents the sum of costs of the states whose action (knowing that the cost of state increases while the position of state increase), $\displaystyle\sum_{i=1}^{N} \mathbb{1}_{\{a_i=r, s_i \in F_1,...,F_f\}}$ is a number of states chosen a preventive replacement action and $\displaystyle\sum_{i=1}^{N} \mathbb{1}_{\{a_i=r, s_i=D\}}$ is a number of states chosen a corrective replacement action.

- $\gamma \in]0, 1[$ represents the discount factor.

5   Numerical Investigation

In this section we compute the optimal replacement policies for various scenarios. For ease of representation and without loss of generality, We will restrict our experiments to industrial components with three health states (H, F and D). We First consider a single monitored component. The optimal maintenance policy is computed for the following parameters: $C_f = 10$, $C_d = 20$, $C_{pr} = 40$ $C_{cr} = 60$ $C_{max} = 100$ $C_o = 800$. The optimal replacement policy is depicted in Fig. 3.

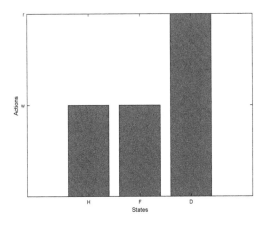

Fig. 3. Optimal policy for a single component with three states.

Next, we illustrate the optimal maintenance policies for two components with three health state. We investigate theses policies depending on whether maintenance grouping is realized or not in Fig. 5 and Fig. 4 respectively. We notice that maintenance are delayed for both components until the other reaches the F or D state to share the associated costs when grouping is allowed.

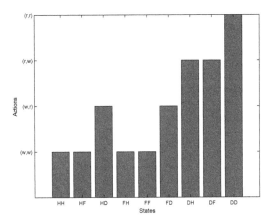

Fig. 4. Optimal policy for two components with three states without maintenance grouping.

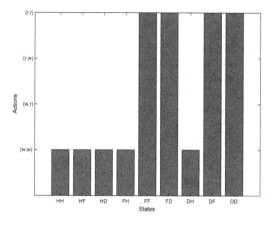

Fig. 5. Optimal policy for two components with three states without maintenance grouping.

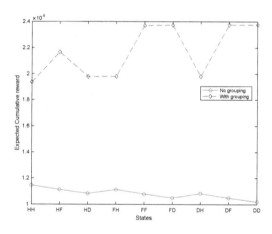

Fig. 6. Expected cumulative reward with and without maintenance grouping.

Figure 6 depicts the expected cumulative obtained by policy iteration algorithm to solve the discounted MDP Γ_2 with three states per component. We notice that grouping maintenance operation increases dramatically the expected reward per state. This is due to maintenance cost sharing and reaches for instance 55% for the state (DD).

6   Conclusion and Perspectives

In this paper we considered the problem of optimal maintenance policy selection for industrial components prone to failure. We modeled the problem using MDP and proved that threshold type optimal policies exist. In a near future, we plan to take into account the cooperation/competition and economic dependence of the components through the use of adapted models.

References

1. Vichare, N.M., Pecht, M.G.: Prognostics and health management of electronics. IEEE Trans. Compon. Packag. Technol. **29**(1), 222–229 (2006)
2. Orsagh, R., Brown, D., Roemer, M., Dabnev, T., Hess, A.: Prognostic health management for avionics system power supplies. In: IEEE Aerospace Conference, pp. 3585–3591 (2005)
3. Scanff, E., Feldman, K., Ghelam, S., Sandborn, P., Glade, M., Foucher, B.: Life cycle cost impact of using prognostic health management (PHM) for helicopter avionics. Microelectron. Reliab. **47**(12), 1857–1864 (2007)
4. Brotherton, T., Jahns, G., Jacobs, J., Wroblewski, D.: Prognosis of faults in gas turbine engines. In: Proceedings of IEEE Aerospace Conference Proceedings, vol. 6, pp. 163–171 (2000)
5. Kirkland, L.V., Pombo, T., Nelson, K., Berghout, F.: Avionics health management: searching for the prognostics grail. In: Proceedings of IEEE Aerospace Conference, vol. 5, pp. 3448–3454 (2004)
6. Wilkinson, C., Humphrey, D., Vermeire, B., Houston, J.: Prognostic and health management for avionics. In: Proceedings of IEEE Aerospace Conference, vol. 5, pp. 3435–3447 (2004)
7. Gertsbakh, I.B.: Models of preventive maintenance (1977)
8. Jardine, A.K., Lin, D., Banjevic, D.: A review on machinery diagnostics and prognostics implementing condition-based maintenance. Mech. Syst. Sig. Process. **20**(7), 1483–1510 (2006)
9. Peng, Y., Dong, M., Zuo, M.J.: Current status of machine prognostics in condition-based maintenance: a review. Int. J. Adv. Manuf. Technol. **50**(1–4), 297–313 (2010)
10. Jeschke, S., Brecher, C., Meisen, T., Özdemir, D., Eschert, T.: Industrial internet of things and cyber manufacturing systems. In: Jeschke, S., Brecher, C., Song, H., Rawat, D.B. (eds.) Industrial Internet of Things. SSWT, pp. 3–19. Springer, Cham (2017). https://doi.org/10.1007/978-3-319-42559-7_1
11. Sadeghi, A.-R., Wachsmann, C., Waidner, M.: Security and privacy challenges in industrial internet of things. In: 2015 52nd ACM/EDAC/IEEE Proceedings of Design Automation Conference (DAC), pp. 1–6 (2015)
12. Hossain, M.S., Muhammad, G.: Cloud-assisted industrial internet of things (IIOT)-enabled framework for health monitoring. Comput. Netw. **101**, 192–202 (2016)
13. Serpanos, D., Wolf, M.: Industrial internet of things. In: Internet-of-Things (IoT) Systems, pp. 37–54 (2018)
14. Kwon, D., Hodkiewicz, M.R., Fan, J., Shibutani, T., Pecht, M.G.: IoT-based prognostics and systems health management for industrial applications. IEEE Access **4**, 3659–3670 (2016)
15. Lee, J., Kao, H.-A., Yang, S.: Service innovation and smart analytics for industry 4.0 and big data environment. Proc. CIRP **16**, 3–8 (2014)
16. Sonntag, D., Zillner, S., van der Smagt, P., Lörincz, A.: Overview of the CPS for smart factories project: deep learning, knowledge acquisition, anomaly detection and intelligent user interfaces. In: Jeschke, S., Brecher, C., Song, H., Rawat, D.B. (eds.) Industrial Internet of Things. SSWT, pp. 487–504. Springer, Cham (2017). https://doi.org/10.1007/978-3-319-42559-7_19
17. Xanthopoulos, A., Kiatipis, A., Koulouriotis, D., Stieger, S.: Reinforcement learning-based and parametric production-maintenance control policies for a deteriorating manufacturing system. IEEE Access **6**, 576–588 (2017)
18. Wan, J., Tang, S., Li, D., Wang, S., Liu, C., Abbas, H., Vasilakos, A.V.: A manufacturing big data solution for active preventive maintenance. IEEE Trans. Ind. Inform. **13**(4), 2039–2047 (2017)

19. Yan, H., Wan, J., Zhang, C., Tang, S., Hua, Q., Wang, Z.: Industrial big data analytics for prediction of remaining useful life based on deep learning. IEEE Access **6**, 17190–17197 (2018)
20. Feng, Q., Bi, X., Zhao, X., Chen, Y., Sun, B.: Heuristic hybrid game approach for fleet condition-based maintenance planning. Reliab. Eng. Syst. Saf. **157**, 166–176 (2017)
21. Fathi Aghdam, F., Liao, H.: Prognostics-based two-operator competition in proactive replacement and service parts procurement. Eng. Econ. **59**(4), 282–306 (2014)
22. Batzel, T.D., Swanson, D.C.: Prognostic health management of aircraft power generators. IEEE Trans. Aerosp. Electron. Syst. **45**(2), 473–482 (2009)
23. Lee, J., Wu, F., Zhao, W., Ghaffari, M., Liao, L., Siegel, D.: Prognostics and health management design for rotary machinery systems - reviews, methodology and applications. Mech. Syst. Sig. Process. **42**, 314–334 (2014)
24. Puterman, M.L.: Markov Decision Process Discrete: Stochastic Dynamic Programming. Wiley, Hoboken (2014)
25. Krishnamurthy, V.: Structural Results for Partially Observed Markov Decision Processes (2015)
26. Sharma, N., Mastronarde, N., Chakareski, J.: Structural Properties of Optimal Transmission Policies for Delay-Sensitive Energy Harvesting Wireless Sensors. arXiv preprint arXiv:1803.09778 (2018)

A Wearable IoT-Based Fall Detection System Using Triaxial Accelerometer and Barometric Pressure Sensor

Elahe Radmanesh[1]([⊠]), Mehdi Delrobaei[1], Oussama Habachi[2], Somayyeh Chamani[1], Yannis Pousset[2], and Vahid Meghdadi[2]

[1] Faculty of Electrical Engineering, K. N. Toosi University of Technology, 1631714191 Tehran, Iran
e_radmanesh@email.kntu.ac.ir, delrobaei@kntu.ac.ir, schamani2002@yahoo.com
[2] University Limoges, XLIM, UMR 7252, Limoges, France
{oussama.habachi,vahid.meghdadi}@xlim.fr, pousset@sic.univ-poitiers.fr

Abstract. The aim of this research work is to develop a wearable and IoT-based fall detection system that can potentially be integrated within a smart home or a community health center to improve the quality of life of the elderly. This system would enable caregivers to remotely monitor the activities of their dependents and to immediately be notified of falls as adverse events. The proposed hardware architecture includes a processor, a triaxial accelerometer, a barometric pressure sensor, a Wi-Fi module, and battery packs. This unobtrusive architecture causes no interference with daily living while monitoring the falls. The output of the fall detection algorithm is a two-state flag, transmitted to a remote server in real-time.

Keywords: Fall detection · IoT architecture · Accelerometry · Pressure sensing · Elderly care

1 Introduction

The worldwide population of the elderly aged 60 or over is expected to more than double by 2050 and to more than triple by 2100, increasing globally from 962 million in 2017 to 2.1 billion in 2050 and 3.1 billion in 2100 [1]. According to the WHO report, falls are one the leading external causes of unintentional injuries [2].

Roughly 40% of falls in elderly are fatal [3], and 20%–30% of non-lethal falls can lead to severe injuries such as lacerations, hip fracture, and head traumata [4]. In the most optimistic case, falls increase disability and extend the rehabilitation period. Women are more likely than men to be injured, leading to twice rate of hip fractures [5]. When an elder person falls and becomes unconscious or is unable to move their body, they succumb to the injuries caused by the fall [6]. Half of those who experienced an extended time lying on the ground passed away within 6 months of the fall [7].

© Springer Nature Switzerland AG 2020
O. Habachi et al. (Eds.): UNet 2019, LNCS 12293, pp. 158–170, 2020.
https://doi.org/10.1007/978-3-030-58008-7_13

Fall also occurs frequently in medical health care centers or hospitals and there are several risk factors which could lead to falls in such settings. These include changes in cognitive, visual, musculoskeletal, sensory or cardiovascular systems as well as extrinsic factors normally related to the environment such as trip hazards and poor lighting [8, 9].

The degree of danger from a fall for aging persons is frequently decided by the location of the fall, time of detecting the fall, duration and time of transfer and rescue services. Automatic detection of fall along with the locations help medical staff to be dispatched immediately [10].

A variety of different tools and methods have been globally developed for smart homes to execute remote health monitoring of physically-impaired and elderly people. A review of the current literature indicates that most elderly people would benefit from real-time, unobtrusive monitoring systems – mainly utilizing wireless sensor networks. Such systems keep track of the daily activities of the elderly at home, enabling them to live independently [10–22].

To detect physical activities (including adverse events such as fall), sensors are either allocated in the surroundings or placed on the person's body to continuously gather data. Based on pre-described detection algorithms, movements are detected and classified as normal or abnormal patterns. The common types of sensors to achieve such a task are listed in Table 1.

Wearable sensors and vision-based systems were initially proposed and later two emerging technologies, depth camera- and radar-based systems, were proposed. It is emphasized in the literature that elderly people prefer unobtrusive in-home sensing with no need to wearing noticeable devices, no interference with daily living, no need to learn new technical skills, and above all, no need to capture video [25].

Table 1. The common technologies for real-time fall detection systems.

Sensor type	Description	Ref.
Accelerometer	Simple linear acceleration sensing to detect any abnormal changes in x, y, or z accelerations	[11, 12]
IMU	Inertial measurement units positioned on the body that generate an alert if there is an abrupt change in motion	[13, 14]
Pyroelectric IR	Array of wall-mounted thermal detectors (passive infrared)	[15]
Barometric pressure	Measuring the altitude of different positions on the human body	[16]
EMG	Wearable, wireless, and minimally invasive surface Electromyography-based setup	[17, 18]
Vision	Human movement detection using vision-based cameras	[19, 20, 30]
Sound and vibration	Floor vibration and sound sensing	[34]
Doppler radar	Detecting object's motion by analyzing changes in the frequency of the returned microwave signal	[21, 22]

(*continued*)

Table 1. (*continued*)

Sensor type	Description	Ref.
UWB radar	Either wall- or ceiling-mounting ultra-wideband radar, providing 3D depth	[23, 24]
IOT Architecture	Includes many sensing devices such as audio, ambient, video, and wearables	[26, 32, 33]

The sensor data (mentioned in Table 1) have been generally utilized for monitoring, supporting, detecting, informing, and gait analysis of elderly people [26]. The main data analysis and classification techniques for fall detection and prediction are listed in Table 2.

Table 2. Main methods of analysis and classification for fall detection and prevention.

Method	Reference
Fuzzy logic	[12]
Bayesian filtering	[27, 28]
K-nearest Neighbor	[29, 36, 37]
Support Vector Machine	[30]
Neural Network	[38, 39]
Principle Component Analysis	[35]
Multi-layer perception	[40]

A complete review of fusion algorithms has been presented in [41]. Fusion algorithms are recent advancements in fall detection and fall prevention studies. In fall detection, data related to posture, inactivity, and presence is analyzed. In fall prevention, fusion systems provide data on human balance (such as static and dynamic sway and ground reaction forces) and gait (such as step and stride lengths, step and stride time, and center of pressure frequency) indices to assess falling risk [41].

It is finally noted that any literature review on the fall detection systems would lack a common agreement on the performance of the classification methods. The possible reasons could be: (a) each study provides a different classification approach based on the understanding of the nature of the fall detection/prevention problem, (b) aside from the classification algorithm, other factors would potentially affect the sensitivity and specificity of the classification method; factors such as signal conditioning method and the extracted features.

None of the sensor types proposed in the literature seem to be necessarily better or worse than other sensor types. Hence, the selection of the sensors (or sensor fusion strategy) depends on various parameters including coverage area, obtrusiveness, privacy issues, power limitations, and cost. The most promising solution to solve the fall detection problem seems to be an IoT-based system, because of the high reliability especially for the alert system that should be as independent as possible of the elderly and at the same time low error.

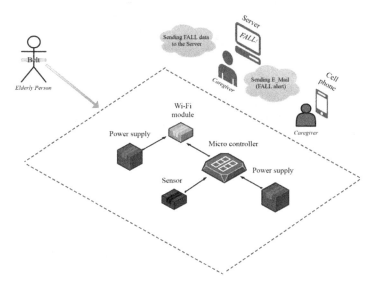

Fig. 1. An overview of the system.

In this paper, we mainly focus on the design preliminary evaluation of a wearable, easy to use, and reliable fall detection system. This system connects to the Internet and transfers a fall-flag to a remote server in real-time (Fig. 1).

2 System Architecture

Figure 1 illustrates an overview of the proposed system and the realized prototype is shown in Fig. 2. The system comprises a microcontroller (main board), a triaxial accelerometer (IMU), a barometric pressure sensor, a Wi-Fi module, and two battery packs (power supply). The sensors are integrated on a single board sized 25 mm by 15 mm (MPU9250, InvenSense Inc., CA and BMP280, Bosch Sensortec, GmbH). The programmable full scale range of the accelerometer is ±2, ±4, ±8 and ±16 g and its normal operating current is roughly 450 μA. BMP280 pressure range is 30,000 Pa to 110,000 Pa with relative accuracy of 12 Pa. The system also employs an STM32 microcontroller (Nucleo-f411RE). The sensors and the microcontroller unit receive data from the sensors via an I^2C bus.

As mentioned in Sect. 1, a variety of different fall detection algorithms have already been developed, but most of them are not optimized to be implemented on a wearable real-time system. In this work, we employed a reliable algorithm, specifically developed and optimized for STM32 microcontrollers (MotionFD middleware library, STMicroelectronics Corp, Geneva, Switzerland) [42].

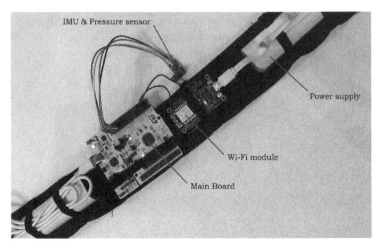

Fig. 2. The hardware architecture of the proposed fall detection system.

This algorithm acquires data from the accelerometers and the barometric pressure data with specific scales and sampling frequencies. The Motion FD API logic sequence and algorithm is illustrated in Fig. 3. According to the method presented in [43], the parameters are calculated as follows:

$$\sigma_{xyz} = \sqrt{\sigma_x^2 + \sigma_y^2 + \sigma_z^2} \tag{1}$$

$$\sigma_a = \sigma\left(\sqrt{A_x^2 + A_y^2 + A_z^2}\right) \tag{2}$$

$$\theta = Arccos\left(\frac{A_z^2}{\sqrt{A_x^2 + A_y^2 + A_z^2}}\right) \tag{3}$$

where σ_{xyz} is the magnitude of the standard deviations of all three axes, σ_a is the standard deviation of the vector magnitude, and θ is the forward falling angle acceleration. The most important difference between σ_a and σ_{xyz} is their degree of dependence to the angle θ; i.e., σ_a is independent of θ, but σ_{xyz} is sensitive to θ.

In order to use this library with the selected sensors, the sample rate of both sensors was set to 25 Hz, the unit of measurements for the accelerometer was set to mg and of the pressure sensor to hPa. The accelerometer data was initially calibrated by subtracting the zero-g-bias voltage on each axis. The zero-g-bias describes the output voltage under solely gravitational forces.

Detecting the falls refers to a specific register configuration (Fig. 4). The acceleration measured along all the axes is set to zero. In a real case, a "fall zone" is defined around the zero-g level (Fig. 4) where all the accelerations are small enough to generate the interrupt. Two important parameters for detecting fall are threshold and duration: the threshold parameter defines the fall zone amplitude while the duration parameter defines the minimum duration of the fall interrupt event to be recognized. The algorithm is able to detect the fall and generate an interrupt if the z-axis acceleration falls below a certain threshold.

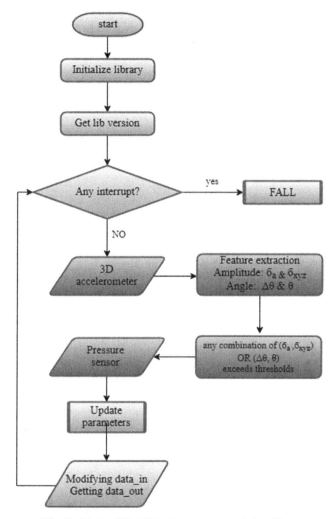

Fig. 3. Motion FD API logic sequence and algorithm.

The suggested settings involve [42] (Fig. 5):

- Input data are the sensor output (select the output data rate ODR $>= 25$ Hz)
- Thresholds can be set to 0.4 g
- Timeout selectable (e.g. $0 \times 03 = 120$ ms which defines the time duration of the fall)
- Output is an interrupt when a fall acceleration profile is detected for 120 ms

There is a register to control the accelerometer data in which four most significant bits (MSBs) are used for adjusting the sensor data frequency as inputs to the library. To modify the time duration, the fifth MSB of FALL register needs to be adjusted. This specific register is also used to configure the threshold parameter (using the rest of the bits). The unsigned threshold values are listed in Table 3. The values given in this table are valid for each accelerometer full-scale range.

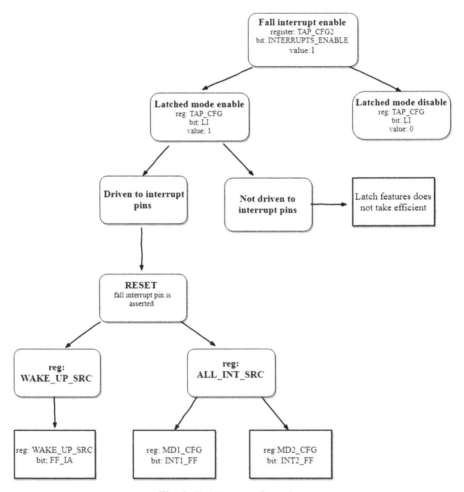

Fig. 4. Registers configuration.

Table 3. Fall threshold LSB value

FALL_FF_THS[2:0]	Threshold LSB value
000	156
001	219
010	250
011	312
100	344
101	406
110	469
111	500

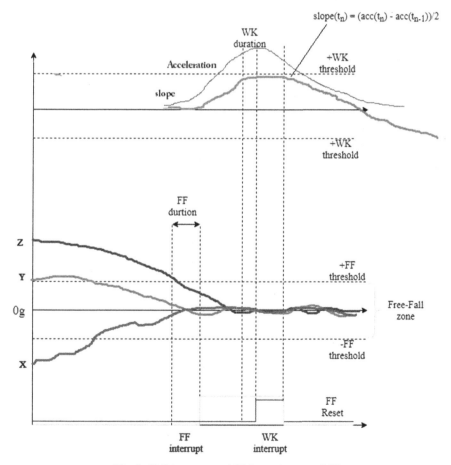

Fig. 5. Fall interrupt and Wake-up interrupt [42].

When two necessary conditions for detecting fall (threshold and time duration) are met, a fall interruption occurs. According to Fig. 4, the tool for resetting the fall interrupt is the WK interrupt. As shown in Fig. 5, the wake-up functionality is based on the comparison of the threshold value with half of the difference between the acceleration of the current and the previous sample data. Therefore, the library will be ready for receiving the next fall interrupt.

Finally, the output of the fall detection algorithm is a two-state flag (either no-fall = 0 or fall = 1). This flag is transmitted in real-time to a server through a low-cost mini WiFi microchip with full TCP/IP stack (ESP8266). The Thinger.io Open Source IoT Platform was also selected to record, manage and display the fall flags in real-time. We can register the data and the time of receiving each data in data buckets and display necessary information on the dashboard section of the server.

3 Results and Discussions

Figure 6 indicates the experimental setup. Once the system is initialized, it begins sending "zeros" to the server which indicates that no fall has initially occurred and the connection with the server is stable. In case a fall event occurs, the system will immediately send a different flag (in the current architecture, a value of 1). The fall event will be displayed and warning messages will be dispatched according to the predefined contacts (either email or text message).

The system was finally tested repeatedly on real (experimental?) falling scenarios at home environment. A single participant performed a variety of different activities of daily living (including sit to stand, body bending, normal walking, and stand to sit). Additionally, forward falls were emulated randomly between the tasks. The participant repeated the tasks (in randomized order) minimum 20 times. None of the normal activities were detected as falls (No false positive). Regarding the falls, only one event (out of twenty) of falls was not detected by the device. The numbers of the errors and the time to display the fall events on the server were recorded. The specifications of the system are listed in Table 4.

Table 4. The specifications of the proposed fall-detection system.

Measure	Value
Net Weight	50 gr
Weight (including batteries and the belt)	300 gr
Processor's power consumption	1.18 W
Transmitter's power consumption	1.44 W
Error rate	$\approx 5\%$
Execution time (on the wearable device)	<1 s
Server display time delay	<60 s

Sit to stand	Walking	Falling	Stand to sit

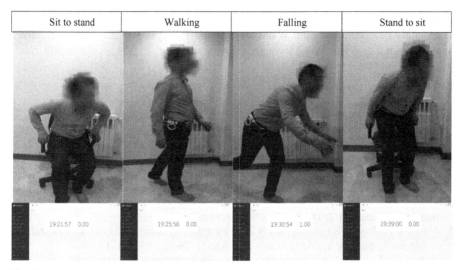

Fig. 6. The experimental setup; the below blocks show the screenshots of the sever corresponding to each incident (including time and the fall flag).

Some current IoT solutions focus on developing algorithms implemented on the server side; hence, sensors' raw data needs to be transferred to the server via the communication link. We proposed a stand-alone Wi-Fi enabled architecture and believe that the proposed architecture is reliable since:

a) An efficient algorithm can easily be implemented on a small-sized processor. The data will be processed locally and the fall-flag will be transferred to the server.
b) The system sends a single flag in real-time, therefore if no data was received from the transmitter, a warning message (inaccessible device) could be immediately generated and sent to the family or health professionals.
c) The modular design of the proposed system allows powering the processor and the transmitter separately. So, if the processor suddenly loses power or the processor fails, this failure (i.e., no flag generated) could be detected locally and an appropriate flag value be transmitted to the server. On the other hand, if the transmitter suddenly loses power, the processor could detect this incident (through monitoring the voltage level of the second battery pack) and enable an appropriate output pin on the processor.

4   Conclusions

Fall detection is one of the major challenges in the elderly care, especially for people living alone. Although many fall detection systems have been proposed in the literature, yet few of them can be easily implemented on a wearable platform. The design of a suitable fall detection system requires an effective integration approach, considering aspects such as hardware architecture, algorithm, and implementation characteristics. We proposed and successfully evaluated a wearable, IoT-based fall detection system that enables caregivers to remotely monitor the activities of their dependents and to immediately be notified of falls as adverse events. The system is simple and reliable, capable of sending notifications through a remote server.

References

1. Unies, Nations: World population prospects: the 2017 revision: key findings and advance tables. UN (2017)
2. World Health Organization: Ageing, and Life Course Unit. WHO global report on falls prevention in older age (2008)
3. Doughty, K., Lewis, R., McIntosh, A.: The design of a practical and reliable fall detector for community and institutional telecare. J. Telemed. Telecare 6(Supplement 1), 150 (2000)
4. Sterling, D., O'Connor, J., Bonadies, J.: Geriatric falls: injury severity is high and disproportionate to mechanism. J. Trauma. 50(1), 116–119 (2001)
5. Vellas, B.J., Wayne, S.J., Romero, L.J., Baumgartner, R.N., Garry. P.J.: Fear of falling and restriction of mobility in elderly fallers. Age Ageing, 26(3), 189–193 (1997)
6. Kaluža, B., Luštrek, M.: Fall detection and activity recognition methods for the confidence project: a survey. In: Proceedings of the 12th International Multiconference Information Society, vol. A, pp. 22–25 (2010)
7. Wild, D., Nayak, U.S., Isaacs, B.: How dangerous are falls in old people at home? Br. Med. J. (Clin. Res. Ed.) 282(6260), 266–268 (1981)
8. King, R.C., Atallah, L., Wong, C., Miskelly, F., Yang, G.Z.: Elderly risk assessment of falls with BSN. In: 2010 International Conference on Body Sensor Networks, pp. 30–35. IEEE. June 2010
9. Mitty, E., Flores, S.: Fall prevention in assisted living: assessment and strategies. Geriatric Nurs. 28(6), 349–357 (2007)
10. Naranjo, P.G.V., Shojafar, M., Mostafaei, H., Pooranian, Z., Baccarelli, E.: P-SEP: a prolong stable election routing algorithm for energy-limited heterogeneous fog-supported wireless sensor networks. J. Supercomput. 73(2), 733–755 (2017)
11. Thammasat, E., Chaicharn, J.: A simply fall-detection algorithm using accelerometers on a smartphone. In: The 5th 2012 Biomedical Engineering International Conference, pp. 1–4. IEEE, December 2012
12. Shen, V.R., Lai, H.Y., Lai, A.F.: The implementation of a smartphone-based fall detection system using a high-level fuzzy Petri net. Appl. Soft Comput. 26(390–400), 2015 (2015)
13. Biroš, O., Karchnak, J., Šimšík, D., Hošovský, A.: Implementation of wearable sensors for fall detection into smart household. In: 2014 IEEE 12th International Symposium on Applied Machine Intelligence and Informatics (SAMI), pp. 19–22. IEEE, January 2014
14. Shastry, M.C., et al.: Context-aware fall detection using inertial sensors and time-of-flight transceivers. In: 2016 38th Annual International Conference of the IEEE Engineering in Medicine and Biology Society (EMBC), pp. 570–573. IEEE, August 2016
15. Sixsmith, A., Johnson, N., Whatmore, R.: Pyroelectric IR sensor arrays for fall detection in the older population. In: Journal de Physique IV (Proceedings), vol. 128, pp. 153–160. EDP Sciences (2005)
16. Lu, W., Wang, C., Stevens, M.C., Redmond, S.J., Lovell, N.H.: Low-power operation of a barometric pressure sensor for use in an automatic fall detector. In: 2016 38th Annual International Conference of the IEEE Engineering in Medicine and Biology Society (EMBC), pp. 2010–2013. IEEE, August 2016
17. Cheng, J., Chen, X., Shen, M.: A framework for daily activity monitoring and fall detection based on surface electromyography and accelerometer signals. IEEE J. Biomed. Health Inform. 17(1), 38–45 (2013)

18. Leone, A., Rescio, G., Caroppo, A., Siciliano, P.: An EMG-based system for pre-impact fall detection. In: 2015 IEEE SENSORS, pp. 1–4. IEEE (2015)
19. Yu, M., Rhuma, A., Naqvi, S.M., Wang, L., Chambers, J.: A posture recognition-based fall detection system for monitoring an elderly person in a smart home environment. IEEE Trans. Inf. Technol. Biomed. **16**(6), 1274–1286 (2012)
20. Harrou, F., Zerrouki, N., Sun, Y., Houacine, A.: Vision-based fall detection system for improving safety of elderly people. IEEE Instrument. Measurement Mag. **20**(6), 49–55 (2017)
21. Liu, L., Popescu, M., Skubic, M., Rantz, M., Cuddihy, P.: An automatic in-home fall detection system using Doppler radar signatures. J. Ambient Intell. Smart Environ. **8**(4), 453–466 (2016)
22. Erol, B., Amin, M.G., Boashash, B.: Range-Doppler radar sensor fusion for fall detection. In: 2017 IEEE Radar Conference (RadarConf), pp. 0819–0824. IEEE, May 2017
23. Diraco, G., Leone, A., Siciliano, P.: Detecting falls and vital signs via radar sensing. In: 2017 IEEE SENSORS, pp. 1–3. IEEE (2017)
24. Diraco, G., Leone, A., Siciliano, P.: A radar-based smart sensor for unobtrusive elderly monitoring in ambient assisted living applications. Biosensors, **7**(4), 55 (2017)
25. Wild, K., Boise, L., Lundell, J., Foucek, A.: Unobtrusive in-home monitoring of cognitive and physical health: reactions and perceptions of older adults. J. Appl. Gerontol. **27**(2), 181–200 (2008)
26. Kumar, E.S., Sachin, P., Vignesh, B.P., Ahmed, M.R.: Architecture for IOT based geriatric care fall detection and prevention. In: 2017 International Conference on Intelligent Computing and Control Systems (ICICCS), pp. 1099–1104. IEEE, June 2017
27. Feng, W., Liu, R., Zhu, M.: Fall detection for elderly person care in a vision-based home surveillance environment using a monocular camera. Signal, Image and Video Processing, **8**(6), 1129–1138 (2014)
28. Liao, Y.T., Huang, C.L., Hsu, S.C.: Slip and fall event detection using Bayesian belief network. Pattern Recogn., **45**(1), 24–32 (2012)
29. Lin, B.S., Su, J.S., Chen, H., Jan, C.Y.: A fall detection system based on human body silhouette. In: 2013 Ninth International Conference on Intelligent Information Hiding and Multimedia Signal Processing, pp. 49–52. IEEE, October 2013
30. Kasturi, S., Jo, K.H.: Human fall classification system for ceiling-mounted kinect depth images. In: 2017 17th International Conference on Control, Automation and Systems (ICCAS), pp. 1346–1349. IEEE, October 2017
31. Han, H., Ma, X., Oyama, K.: Towards detecting and predicting fall events in elderly care using bidirectional electromyographic sensor network. In: 2016 IEEE/ACIS 15th International Conference on Computer and Information Science (ICIS), pp. 1–6). IEEE, October 2016
32. De Luca, G.E., Carnuccio, E.A., Garcia, G.G., Barillaro, S.: IoT fall detection system for the elderly using Intel Galileo development boards generation I. In: IEEE CACIDI 2016-IEEE Conference on Computer Sciences, pp. 1–6. IEEE, November 2016
33. Gia, T.N., et al.: Iot-based fall detection system with energy efficient sensor nodes. In: 2016 IEEE Nordic Circuits and Systems Conference (NORCAS), pp. 1–6. IEEE, November 2016
34. Zigel, Y., Litvak, D., Gannot, I.: A method for automatic fall detection of elderly people using floor vibrations and sound—Proof of concept on human mimicking doll falls. IEEE Trans. Biomed. Eng. **56**(12), 2858–2867 (2009)
35. Hazelhoff, L., Han, J., de With, P.H.N.: Video-based fall detection in the home using principal component analysis. In: Blanc-Talon, J., Bourennane, S., Philips, W., Popescu, D., Scheunders, P. (eds.) ACIVS 2008. LNCS, vol. 5259, pp. 298–309. Springer, Heidelberg (2008). https://doi.org/10.1007/978-3-540-88458-3_27
36. Li, Y., Ho, K.C., Popescu, M.: A microphone array system for automatic fall detection. IEEE Trans. Biomed. Eng. **59**(5), 1291–1301 (2012)
37. Liu, C.L., Lee, C.H., Lin, P.M.: A fall detection system using k-nearest neighbor classifier. Expert Syst. Appl. **37**(10), 7174–7181 (2010)

38. Yuwono, M., Su, S.W., Moulton, B.: Fall detection using a Gaussian distribution of clustered knowledge, augmented radial basis neural-network, and multilayer perceptron. In: 7th International Conference on Broadband Communications and Biomedical Applications, pp. 145–150. IEEE, November 2011

39. Foroughi, H., Aski, B.S., Pourreza, H.: Intelligent video surveillance for monitoring fall detection of elderly in home environments. In: 2008 11th International Conference on Computer and Information Technology, pp. 219–224. IEEE (2008)

40. Tzeng, H.W., Chen, M.Y., Chen, J.Y.: Design of fall detection system with floor pressure and infrared image. In: 2010 International Conference on System Science and Engineering, pp. 131–135. IEEE, July 2010

41. Chaccour, K., Darazi, R., El Hassani, A.H., Andrès, E.: From fall detection to fall prevention: a generic classification of fall-related systems. IEEE Sensors J. **17**(3), 812–822 (2017)

42. STMicroelectronics Homepage. https://www.st.com/en/embedded-software/x-cube-mems1.html. Accessed 25 Apr 2019

43. Tolkiehn, M., Atallah, L., Lo, B., Yang, G.Z.: Direction sensitive fall detection using a triaxial accelerometer and a barometric pressure sensor. In: 2011 Annual International Conference of the IEEE Engineering in Medicine and Biology Society, pp. 369–372. IEEE, August 2011

A Congestion Game Analysis for Route-Parking Selection with Dynamic Pricing Policies

Bassma Jioudi[1,3](✉), Essaid Sabir[2](✉), Fouad Moutaouakkil[1](✉),
and Hicham Medromi[1](✉)

[1] EAS Research Group, ENSEM, Hassan II University of Casablanca, Casablanca, Morocco
b.jioudi@gmail.com, fmoutaouakkil@hotmail.com, hmedromi@yahoo.fr
[2] NEST Research Group, ENSEM, Hassan II University of Casablanca, Casablanca, Morocco
e.sabir@ensem.ac.ma
[3] The Foundation for Research, Development, Innovation and Engineering Sciences (FRDISI),
Casablanca, Morocco

Abstract. The parking plays a fundamental role in urban transport policy development, as an important factor impacting driver's behavior and a major source of traffic problems. In this paper, we present some novel parking pricing schemes to solve the parking spots scarcity and enhance the traffic condition in urban areas. We capture the traveler behavior in term of joint route and parking selection using a simple congestion game. Each traveler aims to minimize his/her expected travel cost by choosing an optimal strategy (route and parking spot). We show how an efficient pricing can incentivize the travelers to optimally choose their route-parking while reducing the traffic congestion.

Keywords: Route-parking selection · Wardrop equilibrium · Dynamic pricing

1 Introduction

Through the 20th century, the private car has gradually turned into the favorable transportation mode. As result the urban areas suffer from an ever-increasing traffic congestion followed by economic, social and environment detriments. In addition, in areas where the parking land is limited and the parking information is not provided, the situation become more challenging.

Based on previous researchers [1], it is estimated that 8–74% of traffic in central business districts (CBD) consists of drivers cruising for parking and the average searching time for a space is about 3.5 to 14 min. This is justified by traveler's preferences for the free or cheap parking lots. So that they get away from off-street lot to on-street parking lots causing many urban traffic problems. Therefore, parking plays a fundamental role in urban transport policy development, as the first vector of modal choice and a potential means of freeing up public spaces and resources. A common behavioral assumption in transport networks studies is that travelers choose the route that they estimate to generate the minimum total travel cost in term of distance and parking charges. The total travel cost is composed of driving distance cost, accessibility or cruising cost, parking

© Springer Nature Switzerland AG 2020
O. Habachi et al. (Eds.): UNet 2019, LNCS 12293, pp. 171–181, 2020.
https://doi.org/10.1007/978-3-030-58008-7_14

fee and walking distance cost to final destination. Accordingly, the parking information should be provided to drivers in real time so that they adjust their trips before their arrivals. Information, as availability, price and location, can be provided by means of smart devices connected to smart parking platform [2–6].

A coherent pricing policy should be defined effectively to balance travel demand. In our work, we focus on the parking price as an important factor that impact the driver behavior and parking demand between off-street and on-street parking lots to maintain the dynamic aspect for urban areas. The regulation of on-street lots price is not necessarily economic in the sense that the authorities seek to maximize the social well-fare. Therefore, the on-street parking lot price should be defined depending on its accessibility which can expressed by the cruising time to find an available lot and the degree of inconvenience caused by the use of the lot for parking on the road.

In this paper we present the impact of parking price on traveler behavior which is described through a routing game. Each traveler aims to minimize their travel cost. The equilibrium resulting from this game is *Wardrop* equilibrium [7] which describes the behavior of travelers when choosing their routes from an origin to a final destination on optimal way. By inducing the parking price, we show how a pricing can entice travelers to choose their routes in the optimal way that minimizes their total travel cost and enhance the congestion traffic. We present the new equilibrium scheme, Optimal equilibrium, resulting from varying the parking price dynamically depending on the time of day. The paper is organized as follow: Sect. 1 list some of the works done in this area. Section 2 presents an on-street parking queuing model were parkers arrive according to discrete Batch Markovian Arrival process (D-BMAP). In Sect. 3, the *Wardrop* and optimal equilibrium are formulated and the simulations are made to show the impact of parking price on traveler behavior and traffic congestion. Finally, we conclude this work.

Parking pricing has attracted lot of researchers as an important element of parking policy [1, 4, 8–11]. It ensures, through a financial constraint for the driver, compliance with the time limit on parking and also provides a source of profit for the parking owner or authorities. The cities of San Francisco [9], Los Angeles [4] and the municipality of Montreal have set up pricing policies in order to solve traffic problems in the cities. The idea is to adjust the parking price dynamically according to real time parking demand (the amount of parking spaces available per parking area) and to communicate the information through the smart devices. Instead of going around in circles, drivers will be able to see where the free parking lots are and at what price; The city of Casablanca in Morocco, where we intend to test and implement our final solution, is implementing a new parking policy, which focus on downtown zones and take place in three operational phases covering the period 2017–2023. Another parallel phase, covering the same period focus on districts out of town. The objectives of these policies are the renew of old parking meters on areas currently managed by parking meter, regulate additional places and construct a new underground parking lot and relay parking.

This paper focuses on the impact of parking price on the driver behavior. By changing the parking price, we simulate the driver choice for routes and parking lot.

2 On-Street Parking Queuing Model

In this section, we model the behavior of drivers looking for an available on-street parking lot. We assume that all parking spaces are occupied and that drivers are looking for parking spaces. Vehicles cruising for a parking lot are supposed to move through the streets in search of the first available parking lot. As soon as a lot becomes vacant, the next cruising vehicle occupies this lot immediately. Drivers looking for parking can be modeled as $D-BMAP/G/1/N-SIRO$, where the queue service is the act of parking, the queue length is the number of drivers cruising and the queue discipline is SIRO - Service in Random Order. We assume that the cruising time is a random variable W. Drivers that are looking for vacated parking lot arrive in Batches of different size according to the Discrete Batch Markovian arrival process *(D-BMAP)* and are selected for service (parking) in random order (SIRO).

The analysis of *D-BMAP* queue have been studied in the literature [12, 13]. In *D-BMAP*, the arrivals are governed by an underlying m-state Markov chain having probability d_{ij}^n, $1 \leq i, j \leq m$, $n \geq 0$, with a transition from state i to state j and a batch arrival of size n. Here $D_0 = (d_{ij}^0)$ is an $m \times m$ non-negative matrix which governs no arrival, whereas $D_n = \left(d_{ij}^n \right)$, $n \geq 1$ are $m \times m$ non-negative matrices which govern a batch arrival of size n. The matrix $D = (d_{ij}) = \sum_{k=0}^{\infty} D_k$ with $De = \mathbf{e}$, where e is a column vector of ones with an appropriate dimension, is a stochastic matrix corresponding to an irreducible Markov chain underlying *D-BMAP*. We call the actual state of this chain the "phase" of the arrival process. Let $\bar{\pi}$ be the $1 \times m$ stationary vector of the underlying Markov chain (UMC), i.e., $\bar{\pi} D = \bar{\pi}$; $\bar{\pi} e = 1$. The fundamental arrival rate (the average number of customer arrivals per slot) of this process is given by

$$\lambda^* = \bar{\pi} \sum_{k=0}^{\infty} k D_k e.$$

Here, we consider a random service where the selection of a customer does not depend on his position in the queue as well as his position in the arrived batch. In this discipline, any waiting customer can be selected for service with equal probability. That is, if there are n (≥ 1) number of customers in the system at a service completion epoch then each customer has an equal chance to get selected for service with probability $\frac{1}{n}$. In addition to that in a batch arrival queueing system, l (≥ 1) number of customers may arrive in the system during the service time of a customer. Hence, a customer is selected for the next service with probability $1/(n+l)$. This process continues, i.e., a customer is selected randomly for service and during the service time, new batches of variable sizes may arrive in the system.

In such a way, we can calculate the cruising time as following:

$\frac{\bar{\pi} n D_n e}{\lambda^*}$, ($n = 1, 2, 3 \ldots$): Is the probability that an arbitrary driver belongs to a batch of size n.

$\frac{d_{ij}^n}{\bar{\pi} D_n e}$, ($n = 1, 2, 3 \ldots$): Is the probability that a batch arrival consisting of n drivers occurs and phase changes from state i to j.

The probability that a driver will occupy a position, say, k^{th} ($1 < k < n$) in a batch of size n is $1/n$.

The state in referred to the state of the chain of parker queue.

The probability that a driver is selected for service, in a recently vacated parking space, depends on the average parking time in this area i.e., a parker leaves after a random parking time. Consequently, the cruising time can be calculated as:

$$W = \frac{1}{\lambda * (1 - PBL_A)} \times \left(\sum_{k=0}^{\infty} \sum_{n=1}^{\infty} \frac{1}{(k+n)} \pi(k) D_n \xi(\mu) \right)$$

Where:

$\pi(k)$: is the probability that a batch of size n find k drivers, looking for parking, before their arrival. k is the queue length;

$\xi(\mu)$: is the service time which refers to the average searching time in this area;

PBL_A: is the lost probability.

Not all drivers are served in this queue, as the arrival rate of traveler drivers exceeds the departure rate of parked vehicles and the capacity is limited. So, we allow drivers in the cruising queue to become discouraged, leave the queue and look for more expensive off-street parking if the cost of on-street parking lot, including cruising cost and walking distance, exceed the cost of off-street lot ($C_p^{on-street} > C_p^{off-street}$).

The cost of parking lot can be expressed as:

$$C_p(\lambda) = C_W(\lambda) + Fee + C_{Walk}$$

It is composed of the marginal cost of cruising time (C_W) and walking time (C_{Walk}) and the fee (*Fee*) of parking that parker should pay for a given parking time.

3   Driver Behavior

Consider a central business district to which the travel demand is inelastic, f. We model the traffic assignment problem where drivers select their routes between origins and finals destinations in transportation networks. The selection of route depends on total travel cost including parking charges. Each traveler aims to minimize their total travel cost. A driver travels to parking close to his destination, parks there for a parking time, and then exits. Each destination has a number of an associated set of on street parking lots and off-street parking lots. We note that several destinations may share the same parking areas.

We consider a directed network $G = (O, A)$ where O is the set of nodes and A is the set of links (arcs). A route $r \in R$ is simply a chain of links $a = (i, j) \in A$ between an origin and final destination. We define a set of finals destinations, D, that drivers travel to as well as a set of parking facilities, P. Each final destination $d \in D$ has an associated set of parking lots p_d composed of a number of on-street and off-street lots $lot_i^d \in p_d$:

$$N_d = \sum_{i=1} lot_i^d$$

Where, lot_i^d is a dummy variable, it's equal to 1 if $lot_i^d \in p_d$ and equal to 0 otherwise; and N_d is the total number of the associated parking lots p_d.

A traveler starts at a single origin node $o \in O$ and travels to his destination $d \in D$. For each $d \in D$, a flow of demand rate equal to f_d should be routed from the corresponding origin o to this final destination d. For $d \in D$, let R_d be the set of routes connecting the corresponding $o - d$, and let $R := \cup_{d \in D} R_d$. A link flow is a nonnegative vector $L = \{l_{a \in A}\}$ describing the traffic rate in each link. Two types of links are considered in this network, a link set constituting the selected route $l_{a \in r}$ and a link set at the selected parking area $l_{a \in p}$. A route flow (f_r) $r \in R$ that represents the demand, that is, $\sum_{r \in R_d} f_r = f_d$ is the total flow from an origin $o \in O$ to a final destination $d \in D$.

Travelers associated with an origin final destination pair $o - d$, can choose their routes among a set of strategies S_o^d. Each strategy defines the route (r) and the parking area (p) selected:

$$S_o^d = \{(p, r) | r \in R_d, p \in p_d\}$$

For the next applications, we need the following notations:

$f_{op}^d = \sum_{r \in R_{od}} (f_{op}^d)^r$: the flow of $o - d$ travelers choosing the parking p;

$f_p^d = \sum_o \sum_{r \in R_{od}} (f_{op}^d)^r$: the total flow of travelers heading destination d and choosing parking p_d.

$f_p = \sum_d \sum_o \sum_{r \in R_{od}} (f_{op}^d)^r$: the total travelers flow choosing parking area p_d.

$f = \sum_d \sum_o f_o^d$: the total travelers flow.

We model the parking traffic as a set of traveler's flows starting from different origins and crossing the network to areas near to their final destinations and then extending uniformly into the parking areas associated with their destinations.

The parking traffic flow is part of the total flow of travelers traffic. Such that:

$$(\lambda_{op}^d)^r = (l_{a \in r} + \frac{1}{\sum_{a \in p_d} |l_a|} l_{a \in p})(f_{op}^d)^r$$

$$\lambda = \sum_d \sum_o \sum_{r \in R_{od}} \sum_{p \in p_d} (\lambda_{op}^d)^r$$

The routing decision of a single traveler depends of the route cost and the choice of other travelers. Each traveler seeks to minimize the total travel cost. The situation resulting from these individual decisions is one in which drivers cannot reduce their journey times by unilaterally choosing another route. it is known as the *Wardrop* equilibrium and His first principle reads *"The journey times on all the routes actually used are equal, and less than those which would be experienced by a single vehicle on any unused route"*.

Each route cost is estimated as:

$$C_r(\lambda) = C_{o-d}(\lambda) + C_p(\lambda)$$

Where, $C_{o-d}(\lambda)$ is the cost of traveling form an origin to an area near to the destination. If the traveler uses his own car, the travel cost is estimated as the cost of fuel consumed and the time needed to reach this area; if the traveler use a multimodal transport system that he parks his vehicle in parking and ride and use the public transit, the travel cost is the public transit charges and the time needed to reach the station closest to the destination.

Interpreting *Wardrop*'s first principle as requiring that all flow travels along shortest paths, a flow $(\lambda_{op}^d)^r$ is called a *Wardrop* equilibrium if and only if for all $d \in D$, we have that:

$$C_r((\lambda_{op}^d)^r) = \min_{x\in R_d, p\in p_d}\left\{C_x((\lambda_{op}^d)^x)\right\}$$

Thus, $C_r(\lambda_{op}^d) < C_x(\lambda_{op}^d)$ for all $d \in D$ and routes r, $x \in R_d$ such that $\lambda_{op}^d > 0$.

The probability that a traveler starting from an origin (o) to a final destination (d) trough route (r) and choosing parking (p) can be obtained by the logit-based probability function:

$$\pi_{op}^{dr} = \frac{e^{(-C_p^r)}}{\sum_{x\in R_{od}, j\in p_d} e^{(-C_j^x)}}$$

Thus, the expected minimum cost for each $o - d$ pair:

$$C_p^r = \min_{x\in R_{od}, j\in p_d} C_j^x = -\ln(\sum_{x\in R_{od}, j\in p_d} e^{(-C_j^x)})$$

The expected traveler flow for $o - d$ pair choosing parking (p) through route (r) is estimated as:

$$\lambda_{op}^{dr} = \lambda \times \pi_{op}^{dr}$$

4   Simulation and Numerical Investigation

We consider an example with one origin node and one final destination. The travelers can choose between two routes and tree parking lots, on-street or two off-street parking lots. The example is illustrated in Fig. 1.

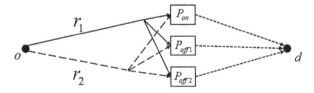

Fig. 1. Example of network with one origin and one destination.

For the simulation we used the data in the Table 1. We simulate the driver behavior in two scenarios with two pricing policies. The first scenario the prices of all parking

Table 1. Data for simulation.

Parameter	Value
The average parking time:	$\mu = 1$ h 45
Total traveler flow:	$\lambda = 400$
The parking price per hour:	$P_{on} = 3$ Dh/h
	$P_{off1} = 5$ Dh/h
	$P_{off1} = 7$ Dh/h
The distance from parking to destination:	50 m $(on-street)$
	400 m $(off-street1)$
	200 m $(off-street2)$
The average walking speed:	4.5 Km/h
The average driving speed:	30 Km/h
The fuel price:	10 Dh/L

lots are fixe but not equal, the price of on-street is the cheapest (3 Dh/h). In the second scenario we use a dynamic pricing policy for on-street lots, so that price vary according to parking occupancy. Each route has a cost depending on the driving distance.

According to simulation results, in the first scenario where the on-street price is fixe (Fig. 2), the total travel cost for travelers choosing on-street lots is generally less than choosing off-street lots. During period between 12 h–15 h the total travel costs are close as the occupancy of on street lot is high which increase the searching cost in this area. For the periods before 8 h and after 18 h. The total travel cost is important for all strategies as the parking occupancy is important due to residents come backing to their home.

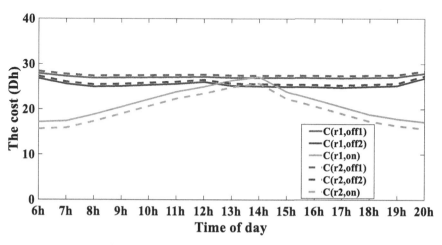

Fig. 2. The total travel cost with on-street fixe pricing policy

Based on the assumption that travelers choose the route that they perceive to generate the minimum travel cost, the Fig. 3 represent the different probabilities of route choice and parking choice. As the cost of on-street lots increase, the probability of choosing these lots decrease and travelers change their preference to off-street lots.

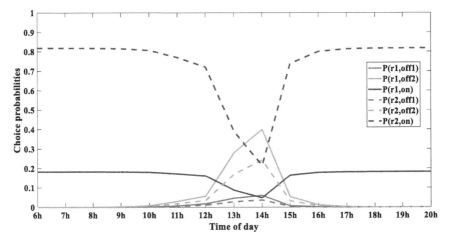

Fig. 3. The choice probabilities with on-street fixe pricing policy.

To balance the parking demand between all parking lots and routes we propose to implement a dynamic pricing policy. So that on-street lots price increases according to occupancy level. Figure 4 represent the variation of the total travel cost by changing the price of on-street lots. All the strategies generate a close cost at 10 h–11 h and 14 h–15 h periods. Due the high occupancy between 11 h–14 h, the on-street strategies generate an important cost that exceed the off-street lots. This can modify the driver's choice expressed in Fig. 5, so that they get away from streets to off-street parking lots.

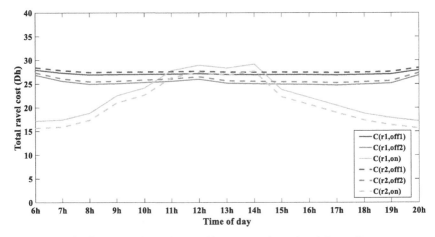

Fig. 4. The total travel cost with on-street dynamic pricing policy.

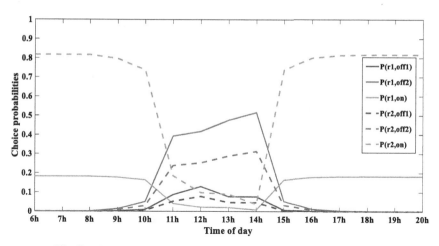

Fig. 5. The choice probabilities with on-street dynamic pricing policy.

Based on routing game theory, the total flow dispersed among all routes so that each traveler selects the route with the minimum travel cost. Figure 6 and Fig. 7 express the dispersion of travelers among routes and parking lots.

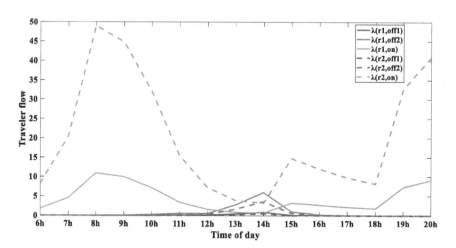

Fig. 6. The travel flow distribution with on-street fixe pricing policy.

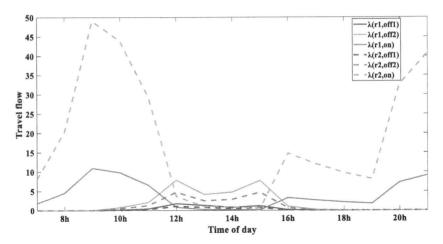

Fig. 7. The travel flow distribution with on-street fixe pricing policy.

5 Conclusion

In this paper, we deal with the impact of parking price on traveler behavior, which is described through a routing game. Each traveler aims to minimize his/her total travel cost. By inducing the parking price, we show that a coherent pricing policy can incentivize the travelers to choose their routes and parking spots that minimize their respective total travel cost and reduce the congestion traffic. As a future direction, and within an areas served by public transportation facilities, we are keen to analyze the impact of multimodal systems on route selection where parking relays are provided free of charge for transiting travelers.

References

1. Cruising for parking - ScienceDirect. https://www.sciencedirect.com/science/article/pii/S09 67070X06000448
2. A cloud-based smart-parking system based on internet-of-things technologies. IEEE J. Mag. https://ieeexplore.ieee.org/abstract/document/7247632
3. Mainetti, L., Marasovic, I., Patrono, L., Solic, P., Stefanizzi, M.L., Vergallo, R.: A Novel IoT-aware Smart Parking System based on the integration of RFID and WSN technologies. Int. J. RF Technol. 7(4), 175–199 (2016)
4. Kotb, A.O., Shen, Y., Zhu, X., Huang, Y.: iParker—A new smart car-parking system based on dynamic resource allocation and pricing. IEEE Trans. Intell. Transp. Syst. 17(9), 2637–2647 (2016)
5. Kianpisheh, A.: Smart Parking System (SPS) Architecture Using Ultrasonic Detector (2012)
6. Lee, C., Han, Y., Jeon, S., Seo, D., Jung, I.: Smart parking system for Internet of Things. In: 2016 IEEE International Conference on Consumer Electronics (ICCE), pp. 263–264 (2016)
7. Wardrop Equilibria - Correa - 2011 - Major Reference Works - Wiley Online Library. https://onlinelibrary.wiley.com/doi/abs/10.1002/9780470400531.eorms0962
8. Larson, R.C., Sasanuma, K.: Congestion Pricing: A Parking Queue Model. Massachusetts Institute of Technology. Engineering Systems Division, Working Paper, August 2007

9. Is the curb 80% full or 20% empty? Assessing the impacts of San Francisco's parking pricing experiment - ScienceDirect. https://www.sciencedirect.com/science/article/pii/S09658 56414000470
10. Optimal occupancy-driven parking pricing under demand uncertainties and traveler heterogeneity: A stochastic control approach - ScienceDirect. https://www.sciencedirect.com/science/article/pii/S0191261514000368
11. Ayala, D., Wolfson, O., Xu, B., DasGupta, B., Lin, J.: Pricing of parking for congestion reduction. In: Proceedings of the 20th International Conference on Advances in Geographic Information Systems, New York, NY, USA, pp. 43–51 (2012)
12. A Batch-Service Queueing Model with a Discrete Batch Markovian Arrival Process—SpringerLink. https://link.springer.com/chapter/10.1007/978-3-642-13568-2_1
13. Analyzing discrete-time D-BMAP/G/1/N queue with single and multiple vacations - ScienceDirect. https://www.sciencedirect.com/science/article/pii/S0377221706008654
14. Lee, H.W., Moon, J.M., Kim, B.K., Park, J.G., Lee, S.W.: A simple eigenvalue method for low-order D-BMAP/G/1 queues. Appl. Math. Model. **29**(3), 277–288 (2005)
15. Hofkens, T., Spaey, K., Blondia, C.: Transient analysis of the D-BMAP/G/1 queue with an application to the dimensioning of a playout buffer for VBR video. In: Networking 2004, pp. 1338–1343 (2004)

Initial Centroid Selection Method
for an Enhanced K-means Clustering Algorithm

Youssef Aamer[1]($\boxtimes$), Yahya Benkaouz[2]($\boxtimes$), Mohammed Ouzzif[1]($\boxtimes$),
and Khalid Bouragba[1]($\boxtimes$)

[1] Telecommunications and Multimedia Group ENSEM, Hassan II University Casablanca,
Casablanca, Morocco
youssef.aamer@gmail.com, ouzzif@gmail.com, bouragba2008@gmail.com
[2] Conceptions and Systems Laboratory FSR, Mohammed V University Rabat, Rabat, Morocco
y.benkaouz@um5s.net.ma

Abstract. Clustering is an important method to discover structures and patterns in high-dimensional data and group similar ones together. K-means is one of the most popular clustering algorithms. K-means groups observations by minimizing distances between them and maximizing group distances. One of the primordial steps in this algorithm is centroid selection, in which k initial centroids are estimated either randomly, calculated, or given by the user. Existing k-means algorithms uses the 'k-means++' option for this selection. In this paper, we suggest an enhanced version of k-means clustering that minimize the runtime of the algorithm using 'Ndarray' option. Experiments have shown that if the first choice of centroids is close to the final centers, the results will be quickly found. Thus, we propose a new concept that provides one of the best choices of starting centroids that reduces the execution time by ≈80% on average for UCI, Shape and Miscellaneous datasets.

Keywords: K-means · K-means++ · Clustering · Initial centroid selection · Performance and runtime

1 Introduction

Clustering is an important method to discover structures and patterns in high-dimensional data. It is a grouping method of a set of objects into disjoint clusters (group of similar objects). This process is used for multiple purposes such as: summary generation [5], customer segmentation [3], fraud detection or as a pre-processing step for forecasting or other data mining tasks. For instance, this approach can be used to find and analyze valuable information related to a specific political party by clustering Facebook users according to their common friends and interests. Clustering techniques help to understand relations between profiles and create a global picture of their traits, and eventually conclude how politicians can have impact on them [4].

Clustering is a method of grouping or partitioning a set of patterns into disjoint clusters (group of data points or objects in a dataset that are similar to other objects in the group and dissimilar to data points in other clusters). It helps to identify similar patterns in the same cluster.

© Springer Nature Switzerland AG 2020
O. Habachi et al. (Eds.): UNet 2019, LNCS 12293, pp. 182–190, 2020.
https://doi.org/10.1007/978-3-030-58008-7_15

There are different categories of clustering algorithms: hierarchical clustering [10] which is an alternative approach to partitioning clustering for identifying groups in the dataset. It does not require pre-specifying the number of clusters to be generated. The result of hierarchical clustering is a tree-based representation of the objects, which is also known as dendrogram. The second type is Fuzzy clustering [12]. It is also known as soft method. It is a standard clustering approach that produces partitions (k-means, PAM), in which each observation belongs to one cluster only. This is known as hard clustering, in Fuzzy clustering. Items can be a member of more than one cluster. Each item has a set of membership coefficients corresponding to the degree of being in a given cluster; the Fuzzy c-means method is the most popular fuzzy clustering algorithm. Density-based clustering [13], is the third category that can find clusters from noisy data using partitioning method which is derived from human intuitive clustering method.

The data is viewed as distributed in the fourth type, which is Model-based clustering [14]. That finds the best models fit to data and estimates the number of clusters. The last category is partitioning algorithms that subdivide datasets into a set of k-groups. The most popular one is k-means, where every cluster is represented by the center of data points that belongs to it.

K-means [16, 17] is an unsupervised algorithm that separates data into k non-overlapping clusters without any label or cluster-internal structure. Objects belonging to the same cluster are similar while objects from different clusters are dissimilar [2]. Though the objective of k-means is to form clusters in such a way that similar samples go into a cluster, and dissimilar samples fall into different clusters. The first step in k-means is to choose centroids, it is considered as the main factor that determines the accuracy of the result and the runtime [18, 19]. There are three options to choose centroids. The first option is to choose centroids randomly. The second option is k-means++ [15] (Implemented by default in the existed API) in which the initial cluster centroids are chosen in a probabilistic way to speed up convergence. The last option consists in choosing the initial centroids that should have the shape of data points. This is the option that we used in our concept. The experiments have proved that if the first choice of the centroids is close to final clusters centers, the algorithm converges rapidly. In this way, we suggest a new concept that provides the best choice of starting centroids, which allows reducing the runtime,

This paper is structured as follows: Sect. 2 describes related works. Section 3 details the suggested concept. The experimental results will be discussed in Sect. 4. Finally, we conclude in Sect. 5.

2 Related Works

To enhance k-means, multiple works have been proposed in the literature. In [5], Xie et al. suggested adding one centroid at each stage of the incremental method they proposed. It starts by designing the centroid of the whole dataset as initial centroid. Then, it computes two clusters where the second centroid is a data point in the dataset. This operation is repeated for all clusters. The runtime for this approach is highly related to the number of clusters, which makes this approach more suitable for small datasets.

In [7], the authors presented the idea of selecting initial centroids based on the average attribute values of dataset elements. Then, search the initial centroids in such a way that the distance is maximum from other selected initial centroids. Finally, repeat the last step, until getting k initial centroids.

David and Sergei propose in [15] a probabilistic method for initial centroid selection that ensures a high accuracy. The suggested selection algorithm is logarithmic. But, they have ignored the huge number of iterations of k-means algorithm to reach stability with a higher clusters presentation quality, which increases the execution time, when they apply k-means on the selection result. This method is implemented in the well-known k-means++. It is the approach against which our approach has been compared in this work.

3   The Suggested Approach

As discussed previously, K-means clustering algorithm is based on an iterative refinement approach [8]. The algorithm starts with initial estimates for the k centroids. These centroids are randomly selected, calculated based on a probabilistic approach or given by the user. Then, the algorithm iterates between two steps. In the first step, each element is assigned to its nearest centroid based on a given distance. In the second step, the centroids are recomputed as the average of all data points in the same cluster. The algorithm terminates when there is no update in the created clusters (stability reached). It has been seen that initial centroids selection has an impact on k-means performances in terms of clusters structure quality and convergence runtime [7].

In this work, we suggest an enhanced k-means algorithms based on a new initial centroids selection approach. The idea is to select the farthest centroids in such a way that they are different and belong to different clusters. This allows maximizing the chance to have data points of different centroids in the same cluster. Therefore, reducing the number of iterations to reach the clusters stability. Consequently, the execution time will be significantly reduced. Figure 1 explains in details the suggested approach.

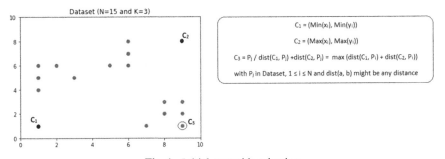

Fig. 1. Initial centroids selection

Figure 1 shows an example that describes how centroids are selected. We assume that we have a dataset of 15 data points. The aim is to partition such dataset into k = 3 clusters. To select the first centroid C_1, we consider a point with the minimum of each dimension, i.e. $C_1 = (Min(x_i), Min(y_i))$ with $1 \leq i \leq N$. Whereas, for the second centroid, we consider the maximum of each dimension, i.e. $C_2 = (Max(x_i), Max(y_i))$.

After the computation of the first two centroids, we calculate the third one in such a way that it is the data point located as far as possible from C_1 and C_2. Therefore, for each data point, we compute the distance from C_1 and C_2. Then we sum up these distances and select the data point which has the maximum of the obtained values. Formally,

$$C_3 = P_j / \, dist(C_1, \, P_j) + dist(C_2, \, P_j) = max \, (dist(C_1, \, P_i) + dist(C_2, \, P_i)) \text{ for all i in } [1..N].$$

This method is general as dist(a, b) might be any distance method such as the Euclidean distance.

Therefore, after applying k-means on this example using our approach to generate initial centroids, we found that we needed just two iterations performed to reach stability. Indeed, stability was reached after the first iteration, while the second one was only performed for confirmation. Consequently, the execution time has been significantly reduced.

Algorithm 1 Enhanced method of selecting initial centroids

Precondition: N is not null and $k >= 2$

1: **function** $bestInitialCentroids(N, k)$
2:   $C \leftarrow null$ ▷ C: a set of centroids
3:   $centroid1 \leftarrow maxDim(N)$
4:   $centroid2 \leftarrow minDim(N)$
5:   $addCentroid(C, centroid1)$ ▷ add the first centroid to C set
6:   $addCentroid(C, centroid2)$ ▷ add the second centroid to C set
7:   **while** $k > size(C)$ **do**
8:     $newCentroid \leftarrow farthestDataPoint(N, C)$
9:     $addCentroid(C, newCentroid)$
10:   **end while**
11:   **return** C
12: **end function**

Fig. 2. Our method algorithm

Figure 2 illustrates the pseudocode of our algorithm which is a function that provides an enhanced method of selecting initial centroids for k-means clustering. In a nutshell, we consider a set N of n data points. Each data point p is projected in a D dimensional space. k is the number of clusters and C is a set of centroids. We also assume a function maxDim(N), that given a data point characterized by maximum dataset dimensions, and another function minDim(N) which return a data point with a minimum dataset dimensions, that are already defined. We assume also a function farthestDataPoint(N, M) with M a set of data points, that, given the farthest data point in N from M set, and a function addCentroids(C, x) that adds a centroid x to a set of centroids C, are already defined. Initially, we start by finding the minimum and maximum dimensions, to create our two first centroids (line 03, 04). Then, assign them (line 05, 06) to the empty set C initialized previously by null (line 02). After, we iterate on the number of clusters k, if it is greater than the number of centroids found, we continue to compute other centroids by calculating in each iteration the farthest data point in the dataset N from centroids previously found (line 8) and add it to the set C (line 09; 11).

Obviously the complexity of such an approach for a dataset containing n data points is $O(n)$, where the complexity of k-means++ centroids selection algorithm is logarithmic [15], which means that the latest performs better than our suggestion. The average complexity of k-means is $O(k\ n\ T)$, where T is the number of iteration [7]. Therefore, the number of iterations T is the main factor of the comparison between our selection method and the probabilistic selection method.

4   Experimental Setup

This section describes the experimental process of our approach. More specifically, we applied both k-means++ and our method on multiple datasets that are also described in the next subsection. Then we compare the results on the execution time and the clusters quality.

4.1   Datasets

We evaluate the performance of the algorithms on three datasets: Shape, UCI and Miscellaneous:

- Shape dataset [2] contains about 29 000 3D shape models.
- UCI dataset [1] contains 468 models used as a service to the machine learning community.
- Miscellaneous Dataset [9] includes 10 handwriting digits and contains 60 000 477 training patterns and 10 000 test patterns of 784 dimensions.

These datasets are presented in Tables 1, 2 and 3 for Shape, UCI and Miscellaneous datasets, respectively.

Table 1. Shape datasets characteristics

Dataset name	Size	Min	Mean	Max	Std
Aggregation	788	1	12.5	36.55	9.91
Compound	399	1	13.24	42.9	9.9
Flame	240	0.5	9.96	27.8	8.53
Jain	373	0.75	12.57	41.3	11.64
Pathbased	300	1	12.65	33.05	9.59
Spiral	312	1	12.25	31.95	9.31

Table 2. UCI datasets characteristics

Dataset name	Size	Min	Mean	Max	Std
Iris	150	0	2430844	5535128	996661.09
Thyroid	215	0	3392455.97	12140369	997671.71
Wine	178	0	26423.86	65535	14302.05
Breast	699	1	3.12	10	2.88
Glass	214	0	11.26	75.41	22.14
Wdbc	569	0	979796.7	911320000	23067472.93
Letter	20000	0	5925.46	15000	2907.86

Table 3. Miscellaneous datasets characteristics

Dataset name	Size	Min	Mean	Max	Std
t4.8k	8001	2	237.97	80	159.35
ConfLongDemo	164860	−2.54	1.65	5.75	1.17

In the tables above, we illustrate datasets statistics that define how data points are organized inside datasets; they can be dispersed or assembled. This information helps us to identify the most suitable dataset category for the both methods (k-means++ and ours).

4.2 Metrics

In order to compare our algorithm, we have considered two evaluation metrics:

- Quality performance: It measures the clusters presentation quality. It is computed based on Silhouette coefficient algorithm [6]. This algorithm is calculated using the mean intra-cluster distance and the mean nearest-cluster distance for each sample.
- Execution time: It computer the running time spent between the selection of the first centroid and the stability of the resulting clusters.

4.3 Results

Tables 4, 5 and 6 present the complete comparison between our method and k-means++. We note that our concept consistently outperformed k-means++, by completing faster in all types of datasets presented. It reduces the average execution time by ≈86%, ≈89% and ≈81% for UCI, Shape and Miscellaneous datasets, respectively. In addition, we almost kept the same amount of quality performance (tested by Silhouette coefficient algorithm [6]) for UCI datasets and shape, and we increase it by ≈7% for Miscellaneous datasets.

We explain the high execution time gain in all datasets that our method provides by the concept used that reduces the number of iteration of k-means algorithm. Especially on dense datasets, our method works more perfectly. We also observe that our method increases also the quality performance on this type of datasets. On the other hand, we note that our method kept the same amount of quality performance or decrease it by 1% in average on dispersed datasets.

Table 4. Experimental results on Shape datasets

Dataset name	K	Number of dimensions	% Runtime gain	% Silhouette coefficient
Aggregation	7	3	89.38%	−0.12%
Compound	6	3	91.12%	7.65%
Flame	2	3	92.07%	−0.11%
Jain	2	3	88.92%	−0.03%
Pathbased	3	3	92.30%	−0.20%
Spiral	3	3	90.72%	0.04%

Table 5. Experimental results on UCI datasets

Dataset name	K	Number of dimensions	% Runtime gain	% Silhouette coefficient
Iris	3	4	92.76%	−0.77%
Thyroid	2	5	88.50%	0.00%
Wine	3	14	88.22%	−1.31%
Breast	2	9	85.51%	0.00%
Glass	7	9	91.03%	−2.30%
Wdbc	2	32	74.90%	0%
Letter	26	16	90.88%	10.77%

Table 6. Experimental results on UCI datasets

Dataset name	K	Number of dimensions	% Runtime gain	% Silhouette coefficient
t4.8k	6	2	77.24%	0.00%
ConfLongDemo	11	3	90.73%	5.26%

5 Conclusion

This paper showed that the choice of initial centroids is one of the main factors that performs k-means clustering algorithm, whether on reducing execution time or enhancing clusters presentation quality. The suggested method has been compared to the k-means++ selection method and significantly reduce runtime. Our algorithm performs well in dense datasets. However, it does not converge quickly on dispersed datasets. Further work on k-means clustering problem might include a more comprehensive study on more varied datasets. This is especially important considering that this study was more efficient on congregated datasets comparing with dispersed datasets.

References

1. http://archive.ics.uci.edu/ml/
2. Gionis, A., Mannila, H., Tsaparas, P.: Clustering aggregation. ACM Trans. Knowl. Discov. Data (TKDD) **1**(1), 1–30 (2007)
3. Bayer, J.: Customer segmentation in the telecommunication industry. J. Database Mark. Cust. Strategy Manag. **17**(3/4), 247–256 (2010)
4. Catanese, S.A., De Meo, P., Ferrara, E., Fiumara, G., Provetti, A.: Crawling Facebook for social network analysis purposes. In: Proceedings of the International Conference on Web Intelligence, Mining and Semantics, p. 52. ACM (2011)
5. Gong, Y., Liu, X.: Generic text summarization using relevance measure and latent semantic analysis. In: SIGIR 2001: Proceedings of the 24th Annual International ACM SIGIR Conference on Research and Development in Information Retrieval, pp. 19–25 (2001)
6. Rousseeuw, P.J.: Silhouettes: a graphical aid to the interpretation and validation of cluster analysis. Comput. Appl. Math. **20**, 53–65 (1987)
7. Baswade, A.M., Nalwade, P.S.: Selection of initial centroids for k-means algorithm. Int. J. Comput. Sci. Mob. Comput. **2**(7), 161–164 (2013)
8. Kanungo, T., Mount, D.: An efficient k-means clustering algorithm: analysis and implantation. IEEE Trans. PAMI **24**, 881–892 (2004)
9. Arthur, D., Vassilvitskii, S.: Worst-case and smoothed analysis of the ICP algorithm, with an application to the k-means method. In: Symposium on Foundations of Computer Science (2006)
10. Gilpin, S., Davidson, I.: Incorporating SAT solvers into hierarchical clustering algorithms: an efficient and flexible approach. In: Proceedings of the 17th ACM SIGKDD International Conference on Knowledge Discovery and Data Mining, San Diego, California, USA, 21–24 August 2011 (2011)
11. Bradley, P.S., Fayyad, U.M.: Refining initial points for k-means clustering. In: Proceedings of the Fifteenth International Conference on Machine Learning, pp. 91–99. Morgan Kaufmann Publishers Inc., San Francisco (1998)
12. Gustafson, D.E., Kessel, W.C.: Fuzzy clustering with a fuzzy covariance matrix. In: Proceedings of the IEEE Conference on Decision Control, San Diego, CA, pp. 761–766 (1979)
13. Yu, Y., Wang, Q., Kuang, J., He, J.: An on-line density-based clustering algorithm for spatial data stream. Acta Automatica Sinica **38**(6), 1051–1059 (2012)
14. Fraley, C., Raftery, A.: How many clusters? Which clustering method? Answers via model-based cluster analysis. Comput. J. **41**(8), 578–588 (1998)

15. Arthur, D., Vassilvitskii, S.: k-means++: the advantages of careful seeding. In: Proceedings of the Eighteenth Annual ACM-SIAM Symposium on Discrete algorithms, SODA 2007, pp. 1027–1035. Society for Industrial and Applied Mathematics, Philadelphia (2007)
16. Avrithis, Y., Kalantidis, Y., Anagnostopoulos, E., Emiris, I.Z.: Web-scale image clustering revisited. In: Proceedings of the 2015 IEEE International Conference on Computer Vision (ICCV), 07–13 December 2015, pp. 1502–1510 (2015)
17. Elkan, C.: Using the triangle inequality to accelerate k-means. In: Proceedings of the Twentieth International Conference on International Conference on Machine Learning, Washington, DC, USA, 21–24 August 2003, pp. 147–153 (2003)
18. Arai, K., Barakbah, A.R.: Hierarchical K-means: an algorithm for centroids initialization for k-means. Rep. Fac. Sci. Eng. Saga Univ. **36**(1), 25–31 (2007)
19. Fahim, A.M, Salem, A.M, Torkey, F.A., Ramadan, M.A.: An efficient enhanced k-means clustering algorithm. J. Zhejiang Univ. Sci. **7**, 1626–1633 (2006)

Pervasive Services and Applications

Dependency Between the Distance and International Voice Traffic

Zagroz Aziz[✉] and Robert Bestak

Czech Technical Universiy in Prague, Technicka 2, Prague 6, Czech Republic
{azizzagr,robert.bestak}@fel.cvut.cz

Abstract. The mobile voice traffic can provide useful information as for importance and relation of interconnected countries. In this paper, we investigate the international outgoing/incoming voice traffic dependence on the destination, based on the three-month Call Detail Records dataset analyzes. The distance between countries, more precisely the distance between centroids of countries, is calculated by using the great-circle distance approach. Additionally, the voice traffic parameters are normalized with respect to population of countries to obtain comparable outcome independently of the country population. The obtained results demonstrate dependency of traffic parameters on destinations, where the nearby countries are responsible for more than 50% of the total traffic.

Keywords: Mobile network · Voice traffic · Waiting time · Service time · Interarrival time · Centroid · Great-circle distance · Population

1 Introduction

Telecommunication traffic volume between two countries is usually based on the nature of relationship that the two countries have; such as tourism, economic, social aspects, etc. The strength and type of the relationship can be estimated by analyzing the volume and the time profiles of corresponding originating/terminating traffic, including both data and voice traffic.

In this paper, we employ Call Detail Records (CDR) of an international voice traffic of a mobile network carrier to extract voice traffic parameters to analyze the traffic among countries. An international voice traffic carrier network represents a medium between two mobile telcoes located in two different countries. Notice, there can be more than one carrier between the two countries. The voice traffic can be either based on Session Initiation Protocol (SIP) or on Time Division Multiplexing (TDM).

A CDR record captures every phone call activity and logs both national and international. It is created in the core network and provides detailed information about the call scenario setup; e.g., the time and date of call, network elements involved in the connection, the connection type, the calling and called parties numbers, etc. Hereafter, we denote the country which stores the CDRs as the origin country and the opposite side as the destination country.

© Springer Nature Switzerland AG 2020
O. Habachi et al. (Eds.): UNet 2019, LNCS 12293, pp. 193–204, 2020.
https://doi.org/10.1007/978-3-030-58008-7_16

The CDR data can serve for various of purposes; such as network performance indicating, optimizing, planning or dimensioning. Furthermore, the outgoing and incoming voice traffic parameters from CDRs can be extracted, such as waiting time, service time or interarrival time [1]. These parameters can help us in understanding and analyzing the volume of traffic per specific regions or countries [2].

Another important factor influencing voice traffic characteristics between two interconnected countries is the distance among them. Technically speaking, the actual call routing path (number of routing points, distance between two routes, locations of routes, network media, etc.) defines the real distance between two interconnected calls (see Fig. 1). However, such information is hardly available. Therefore, to calculate theoretical distance between two countries, we use in our paper a great-circle distance approach [3], which is specified as the shortest distance between two points on the surface of sphere.

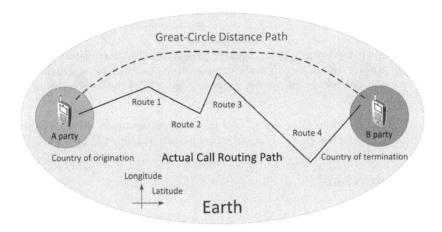

Fig. 1. Great-circle distance vs. actual path routing.

When calculating distance between two countries, we encounter 3 main issues: i) two reference location points have to be assigned ii) a country is not represented as a point but a highly irregular polygon, and iii) the major cities are sporadically located within countries, mainly in case of countries with large areas. To overcome these issues, we consider the country centroid as the reference point, i.e. we assume the distance between two countries to be the distance among their centroids [4] (see Fig. 2).

Another aspect that highly influences the traffic volume is the country population. Naturally, highly populated countries generate higher traffic than less populated countries. To eliminate this population effect, we applied a normalization to adjust values measurements on different scales to a notationally common scale.

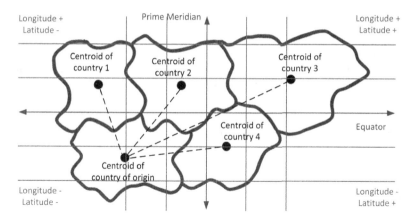

Fig. 2. Mapping centroid of countries on spherical coordinates of earth.

The major contribution of our paper can be summarized as:

– Extracting voice traffic characteristics from the raw CDR dataset,
– Mining voice traffic characteristics to evaluate the amount of generated traffic between the countries,
– Analyzing dependency of the voice traffic parameters based on the destination while taking into account the population of countries.

The rest of the paper is organized as follows. Section 2 outlines related works. The utilized data sources, their processing and merging together are detailed in Sect. 3. Section 4 provides and discusses the obtained results. Finally, our work is summarized in Sect. 4.

2   Related Work

In the last decade, the role of CDRs for different sort of purposes are gradually increasing. They are mainly implemented to analyze and optimize the network performance (radio access or core networks, charging and billing systems, etc.), or to evaluate impacts of implemented company policies.

In [5], authors present a framework to process large and complex raw CDR datasets. The suggested framework is then applied to efficiently manage raw data to help improving communication service provider data management and processing.

In [6], a system is built to analyze mobile network General Packet Radio Service metadata. The authors exert CDR data and Internet Protocol Detail Records to detect potential criminal activities over end-to-end encryption messaging such as Whatsapp, Facebook messenger, etc. The paper concludes that the system reduces time, which is required by authorities for data processing in their investigations.

A data cleaning process of 2.5 million CDRs to filter anomalies are proposed in [7]. The proposal uses social network analysis to analyze behaviors and relationships between customers through their calling profiles. The paper concludes that with the proposed measures, more accurate results can be attained in pre-processing data and detecting anomaly numbers in user profiles.

Authors in [8] utilize a Hadoop based mobile big data processing platform to find out user mobility behaviors. The analyzed data consist of 6 TB, originating from a 4G network. The presented results show that 4G data traffic can provide relatively finer granularity about user mobility and location.

In [9], the authors present CDR dataset of 27604 mobile network base stations from 75 zones in Beijing. The paper proposes the use of CDR data for traffic distribution and location update analysis. According to the analysis, the authors conclude that weekly traffic variation in the studied zones meet the power law distribution.

One-month CDR dataset of 10000 subscribers from a Chinese telecom operator is investigated in [10]. Authors divide abnormality in users' behaviors through the agglomerative hierarchical clustering method. The paper concludes that the anomalous users are mainly distributed in the urban areas.

In [11], authors have developed a scheme to store CDRs in a data warehouse and process them through online analytical processing tools. The application of this schema is illustrated on categorizing user profiles to understand customers' behaviors according to marketing offers (i.e.: offering special rates, deals, bonus, etc.

Authors in [12] identify weekly patterns of human mobility based on classifications of user profiles through 800 million CDRs. The results in this proposal enhance the human mobility investigation by local authorities or to do the urban planning.

In comparison to the previously mentioned works, in our paper, we provide an analytical study based on CDR data from an international voice traffic network carrier. We are interested in finding dependency between voice traffic parameters and phone call destinations by combining data from different data sources.

3   Data Processing

In this section, we provide more details regarding the used data sources (CDRs, countries distance and population datasets), their processing and combination.

3.1   CDR Dataset

In our study, a three-month period of 9 million CDR records in raw format (October-December 2016) of an international outgoing/incoming voice traffic originating from a mobile network carrier are studied. A batch of several CDRs, which form so called a CDR file, is generated and stored in the system every 15 min. The CDR file contains thousands of lines of CDRs, where each line indicates a call status.

In CDRs, there are several parameters that play a vital role in a variety of network analysis and optimization, but we are interested in the traffic evolution aspects per countries. For such investigations, we exploit voice traffic parameters such as waiting, service and interarrival time. These three parameters are essential to analyze the impact of different destinations (i.e., countries) on the voice traffic evolution.

The waiting time represents the time period between the connecting and the connection phase of call. It is usually the sum of ringing and network call processing time. The ringing time period depends on several aspects such as the calling party patience, the called party awareness of the call, time zone difference, etc. On the other hand, the network time period goes through different processing phases before connecting the two parties. It also depends on how the two interconnected networks handle the call (e.g., system configuration, routing path, called party location area, etc.).

The service time is the total duration of the call. It refers generally to the interval between the time period when the phone is picked up (connection time) and hung up (disconnection time). Finally, the interarrival time is the time period between two consecutive call arrivals [13]. The average time between two calls is a key indicator of the traffic amount between two interconnected networks that is used for resource or/and network dimensioning purposes.

Notice that in the preprocessing phase of the CDR dataset, we removed from the dataset countries with negligible number of calls and negligible service time duration. Thus, the results is not influenced by outliers.

3.2 Population Dataset

An important factor that should be taken into account when investigating the traffic is the number of habitants per country. Highly populated countries naturally generate more traffic than the lower populated countries. For instance, a country of 1 million habitants with 1000 voice calls per time unit makes no difference in the voice traffic from a country of 10 million population with 10000 voice calls per time unit. That is, at the first sight both countries would be seen similarly important from the traffic point of view. However, by considering the difference in population of both countries, the first one would be also as important as the second one.

To avoid this population effect and to obtain relative values, we apply a normalization to the voice traffic according to the population datasets. In our study, we use a publicly available UN population dataset [14]. The coefficient of normalization for a country j, denoted c, is used to determine the number of calls per 10^3 inhabitants and we calculate it as:

$$c_j = \frac{n_j}{p_j} \times k \tag{1}$$

where the n_j represents the total number of calls for the given country j per the three-month period, and p_j indicates the population of country j, and the parameter $k = 10^3$. In the equation, we propose $k = 10^3$ as more reasonable results for both populations of countries and number of calls can be obtained. For instance, $k = 10^2$ is relatively small for higher populated countries, whereas $k = 10^4$ is relatively high for the least populated countries in our analyzed dataset.

3.3   Distance Dataset

Geographic coordinate system allows every single location point on earth to be defined by latitude (the north-south position of a location point on the earth), resp. longitude (the east-west position of the same location point). The geographical coordinates are exerted to measure the distance between countries. As stated in Sect. 1, it is not feasible to calculate the distance based on the physical routing path. We use in our case the great-circle distance approach to calculate the distance between countries. Another issue represents the measurement of distance between countries as countries are polygons, and not simple points. To overcome this, we use centroids of countries to calculate distance [15], see Fig. 2.

To determine the great-circle distance between the centroids of origin, i and the destination countries, j, the haversine formula is applied [16]:

$$a_{ij} = \sin^2(\frac{\Delta\varphi_{ij}}{2}) + \cos\varphi_i . \cos\varphi_j . \sin^2(\frac{\Delta\lambda_{ij}}{2}) \tag{2}$$

$$c_{ij} = 2.\arctan 2(\sqrt{a_{ij}}, \sqrt{(1 - a_{ij})}) \tag{3}$$

$$d_{ij} = R.c_{ij} \tag{4}$$

where φ (λ) is the latitude (longitude) of country centroid, Δ is the difference between φ and λ, and d is the distance between centroids. The earth radius, R, is set to 6371 km [17].

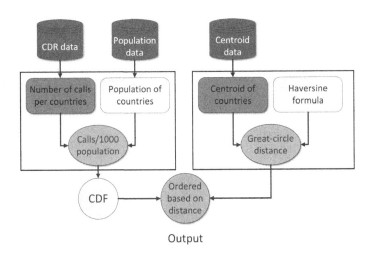

Fig. 3. Merging and processing of data sources.

Notice that the d_{ij} represents the theoretical distance between two countries i and j, as in the reality, the distance is given by the physical communication routing path as explained in Sect. 1.

3.4 Data Merging

The whole data processing is illustrated in Fig. 3. In the first step, the voice traffic data of countries are exploited from the CDR dataset. Secondly, we normalize the population of countries to voice traffic distribution to eliminate the population effect as mentioned in Sect. 1. Furthermore, we employ the centroid data to determine the great-circle distance among countries centroids. Finally, the calculated distance data are ordered from the nearest to the furthest to the country of origin with the normalized data.

4 Results and Discussion

Our analyses are done in two steps. In the first step, we determine and analyze the parameters of outgoing/incoming international voice traffic: i) the waiting time, ii) the service time, and iii) the interarrival time. The mean values of these parameters are shown in Fig. 4 and Fig. 5 in accordance with the total number of calls per 3-month period and per countries. In the figures, we jointly show all three parameters for the outgoing and incoming traffic.

On x-axis, the countries are ordered from the highest number of calls to the lowest ones, i.e. the last country has only a few calls per the 3-month period. On right (resp. left) y-axis, the total number of calls per country (resp. the mean values of waiting, service and interarrival time) are illustrated. From the figures, the mean values of waiting time and service time are nearly independent on the number of calls. Whereas the interarrival time manifests dependency on the number of calls.

When comparing Fig. 4 and Fig. 5, the incoming waiting time shows higher stability over the outgoing one. This is due to the fact that a calling party usually experiences different waiting time according to the destination. In other words, the outgoing traffic fits the habits of each country around the world individually, while the whole world traffic behavior is with respect to the reference network.

In the second step, we demonstrate the influence of destination on the generated outgoing/incoming international voice traffic. In Fig. 6 and Fig. 7, the countries are ordered based on the distance (x-axis); from the nearest to the farthest destinations from the reference country. On the right y-axis, the cumulative distribution function (CDF) of outgoing/incoming traffic ratio is shown. The ratio is calculated by considering the number of calls per 10^3 population unit of countries (as explained in Sect. 3.2). The CDF is applied to investigate the voice traffic distribution among countries based on the ordered distance. On the left y-axis, the distance of countries (in kilometers) to the reference country is indicated.

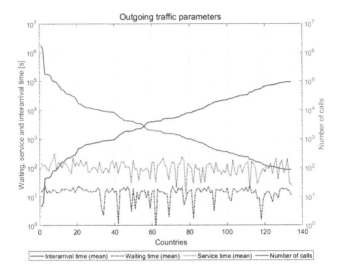

Fig. 4. Outgoing traffic parameters.

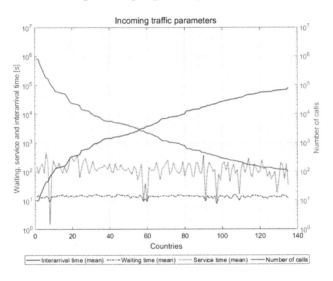

Fig. 5. Incoming traffic parameters.

As can be observed from Fig. 6, there can be distinguished about 3 distance regions, where the traffic importantly grows: a) $0 \div 1500$ km, b) $3000 \div 5000$ km, and c) $5000 \div 6000$ km.

The major outgoing traffic increasing is quite noticeable for the first 15 nearest destinations to the reference one, where most of these destinations are neighboring countries, directly or indirectly connected via their borders; these countries generate more than 60% of the total traffic.

These countries geographically share many cultural behaviors; common relating ethnicities or tribes that only political border separates them. Politically speaking, business, tourism and economic relations are among the most important factors that neighboring countries share. Indirect neighbors are those countries that the neighboring countries lie in between. Those countries are also tightly linked through their business, tourism, or economic sectors.

The countries lying in the second region, $3000 \div 5000$ km, take about 20% of the total traffic. These countries feasibly share historical (e.g., migration) or trading, business and tourism aspects with the reference country.

In the third region, $5000 \div 6000$ km, countries produce about 10% of the total traffic. This portion can be seen as relatively high comparing to more distant countries, but a closer inspection reveals that only few countries contribute to this traffic. The relationship of such countries with the reference country can be either due to tourism or business.

The remaining countries, above 6000 km, contribute to the total traffic in about 5%. Those countries are typically out of major interests for the reference country, where the traffic is mainly due to temporal people visits (e.g., tourism).

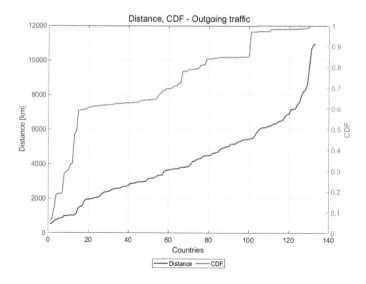

Fig. 6. Outgoing traffic based on ordered distance.

According to the incoming traffic in Fig. 7, we can also distinguish about 3 distance regions, where the traffic importantly grows; a) $0 \div 1500$ km, b) $3000 \div 4000$ km, and c) around 5000 km. In the first region, $0 \div 1500$ km, the major incoming traffic increasing comes from the nearest countries to the reference one, where these countries generate about 50% of the total traffic. As in Fig. 6, these countries have e strong relationship with the reference one (culture, business, tourism, etc.).

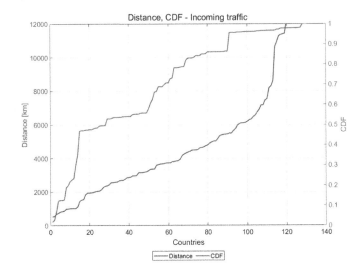

Fig. 7. Incoming traffic based on ordered distance.

The countries lying in the second region, (resp. third region) $3000 \div 4000$ km (resp. around 5000 km), produce about 30% (resp. 10%) of the total traffic. The rest of countries, located above 6000 km, contribute slightly to the total traffic, and this traffic is mainly the result of tourism.

5   Conclusion

In this paper, we study the international voice traffic dependency per destinations. The analysis is done based on three-month CDR dataset, which consist of about 9 million CDRs of outgoing/incoming international voice traffic. The distance between countries, more precisely the distance between countries' centroids, is determined by using the great-circle distance approach. To avoid the population traffic effect, a data normalization is applied to structure the relational voice traffic data to countries populations.

From the CDR data we extract voice traffic parameters such as waiting, service and interarrival time for each country. The obtained results show dependency of voice traffic parameters on the destinations. The nearby countries are responsible for more than 60% (resp. 50%) of outgoing (resp. incoming) traffic. This is mainly due to historical, cultural, business, tourism and economy factors.

In our future work, we plan to investigate in Policy and Charging Rules Function data and to use them to detect network anomalies due to inconsistent configuration of terminal and/or network parameters.

Acknowledgements. This research work was supported by the Grant Agency of the Czech Technical University in Prague, grant no. SGS18/181/OHK3/3T/13.

References

1. Geetha, S., Ramalakshmi, V., Bhuvaneeswari, S., Kumar, Bh.: Evaluation of the performance analysis in fuzzy queueing theory. In: International Conference on Computing Technologies and Intelligent Data Engineering (ICCTIDE 2016), Kovil-patti, India (2016). https://doi.org/10.1109/ICCTIDE.2016.7725348
2. Niknafs, M., Ukhov, I., Eles, P., Peng, Z.: Two-phase interarrival time prediction for runtime resource management. In: Euromicro Conference on Digital System Design (DSD), Vienna, Austria (2017). https://doi.org/10.1109/DSD.2017.42
3. Fukushima, T.: Fast transform from geocentric to geodetic coordinates. J. Geodesy **73**, 603–610 (1999). https://doi.org/10.1007/s001900050
4. Thomas, G., B., Weir, M., D., Haas, J., Heil, Ch.: Thomas' Calculus, 13th edn., Ch. 6, pp. 410 (2014)
5. Ali, A.R.: Real-time big data warehousing and analysis framework. In: IEEE 3rd International Conference on Big Data Analysis (ICBDA), Shanghai, China (2018). https://doi.org/10.1109/ICBDA.2018.8367649
6. Joshi, A., Oberoi, M., Bose, R.: Analyzing CDR/IPDR data to find people network from encrypted messaging services. In: IEEE 4th International Conference on Collaboration and Internet Computing (CIC), Philadelphia, PA, USA (2018). https://doi.org/10.1109/CIC.2018.00013
7. Werayawarangura, N., Pungchaichan, Th., Vateekul, P.: Social network analysis of calling data records for identifying influencers and communities. In: 13th International Joint Conference on Computer Science and Software Engineering (JCSSE), Khon Kaen, Thailand (2016). https://doi.org/10.1109/JCSSE.2016.7748864
8. Sun, W., Miao, D., Qin, X., Wei, G.: Characterizing User Mobility from the View of 4G Cellular Network. In: 17th IEEE International Conference on Mobile Data Management (MDM), Porto, Portugal (2016). https://doi.org/10.1109/MDM.2016.19
9. Wang, X., Dong, H., Zhou, Y., Liu, K., Jia, L., Qin, Y.: Travel distance characteristics analysis using call detail record data. In: 29th Chinese Control and Decision Conference (CCDC), Chongqing, China (2017). https://doi.org/10.1109/CCDC.2017.7979109
10. Wang, Zh., Zhang, S.: CDR based temporal-spatial analysis of anomalous mobile users. In: IEEE 14th Intl Conference on Dependable, Autonomic and Secure Computing, 14th Intl Conference on Pervasive Intelligence and Computing, 2nd Intl Conference on Big Data Intelligence and Computing and Cyber Science and Technology Congress (DASC/PiCom/DataCom/CyberSciTech), Auckland, New Zealand (2016). https://doi.org/10.1109/DASC-PICom-DataCom-CyberSciTec.2016.126
11. Maji, G., Sen, S.: Data warehouse based analysis on CDR to retain and acquire customers by targeted marketing. In: 5th International Conference on Reliability, Infocom Technologies and Optimization (Trends and Future Directions) (ICRITO), Noida, India (2016). https://doi.org/10.1109/ICRITO.2016.7784955
12. Thuillier, E., Moalic, L., Lamrous, S., Caminada, A.: Clustering weekly patterns of human mobility through mobile phone data. IEEE Trans. Mob. Comput. **17**(4), 817–830 (2018). https://doi.org/10.1109/TMC.2017.2742953
13. Feng, H., Shu, Y.: Statistical analysis of packet interarrival times in wireless LAN. In: International Conference on Wireless Communications, Networking and Mobile Computing, Shanghai, China (2007). https://doi.org/10.1109/WICOM.2007.473
14. UN. https://population.un.org/wpp/Download/Standard/Population/. Accessed 11 June 2019

15. Google developers Community. https://developers.google.com/public-data/docs/canonical/countries.csv. Accessed 11 June 2019. The Periscope Data Community. https://community.periscopedata.com/t/63fy7m/country-centroids. Accessed 11 June 2019

16. Chopde, N., R., Nichat, M., K.: Landmark based shortest path detection by using A* and haversine formula. In: International Journal of Innovative Research in Computer and Communication Engineering, vol. 1 (2013). ISSN (Online): 2320–9801

17. Poirier, J. P.: Introduction to the Physics of the earth's interior, 2nd edn., Ch. 7, p. 223 (2003)

IoT Platforms for 5G Network and Practical Considerations: A Survey

Sejuti Banik[1,2], Irvin Steve Cardenas[1], and Jong-Hoon Kim[1(✉)]

[1] Advanced Telerobotics Research Laboratory, Kent State University,
Kent, OH 44242, USA
{icardena,jkim72}@kent.edu
[2] The Department of Electrical and Computer Engineering,
The University of Akron, Akron, OH 44325, USA
sb343@zips.uakron.edu
http://www.atr.cs.kent.edu

Abstract. The fifth generation (5G) mobile network will enable the Internet of Things (IoT) to take a large leap into the age of future computing. As a result of extended connectivity, high speed, and reduced latency services being provided by 5G, IoT has experienced and will continue to undergo a remarkable transition in every field of daily life. Furthermore, fog computing will revolutionize the IoT platforms by decentralizing the operations through the cloud and ensuring sustainability with big data, mobility, and reduced processing lag. 5G is ubiquitous, reliable, scalable, and economic in nature. The features will not only globalize IoT in a broader spectrum, but also make common people interact smartly and efficiently with the environment in real time. In this study, a combined survey is presented on different IoT applications coupled with cloud platforms. Moreover, the capabilities of IoT in the influence of 5G are explored as well as how the IoT platform and services will adopt through 5G are envisaged. Additionally, some open issues triggered by 5G have been introduced to harness the maximum benefit out of this network. Finally, a platform is proposed to implement in the telepresence project based on the investigation and findings.

Keywords: IoT · Fog computing · 5G network · IoT platforms

1 Introduction

The paradigm of the Internet of Things (IoT) will cross new boundaries in the coming decade due to the continuous advancement of technology. In simple words, IoT can be defined as the extension of Internet connection to everyday physical devices enabling, smart interactions between other devices and people. The activities and interactions of such devices, or things, (e.g. sensors and other hardware) in an IoT network can be monitored, and remotely controlled through ubiquitous end-devices such as smart phones or personal computers. Progress in IoT architecture(s) and infrastructure(s) have turned the concepts of smart

O. Habachi et al. (Eds.): UNet 2019, LNCS 12293, pp. 205–225, 2020.
https://doi.org/10.1007/978-3-030-58008-7_17

homes, smart medical care, smart transportation, vehicles-to-everything (V2X) communication, smart environment monitoring, and smart agriculture into tangible implementation.

Contemporary IoT platforms provide services enabling the interaction between hardware devices, software and web services that are part of an IoT system. In specific, these platforms provide the middleware necessary to successfully connect hardware, negotiate between the communication protocols of different hardware and software, access the cloud for storage, perform analysis and processing on IoT data, and also provide services that are aimed at providing privacy and security of communication among devices. Some platforms even provide device-specific support and software (e.g. Arm MBED, RTOS for MCUs) [1]. From a hardware perspective, electronic chips like QCA4020 SoC [2], by Qualcomm Technology, are interoperable with devices that use different radio technologies; as well as integrate with other existing software platforms, IoT ecosystems and cloud services. According to IoT Analytics [3], there are over 450 IoT platforms catering to the changing demands of personal, and commercial usage.

But as the number of users rises, the confined range of current 4G towers incur bandwidth deficiency, and their accommodation capacity is exceeded. Furthermore, an increasing need to process huge amounts of data generated each second by the enormous pool of devices brings an increase to communication latency. The 5G network helps to eliminate these limitations. High speed, capacity for larger number of devices without bandwidth constraint, and reduced latency are few of its advantages.

The overwhelming number of IoT platforms often leads to confusion over which platform suits the intended purpose of the users. Different platforms offer different services which are essential in specific use-cases. Hence, this paper attempts to provide a survey on the unique feature set and limitations of some of the most popular platforms and how they will conform to 5G. Additionally, a platform enabled by 5G is proposed for future application in telepresence robots.

Lastly, a brief overview of the privacy issues, their theoretical solutions and the apparent disadvantages of 5G have been discussed to draw the conclusion.

2   The Paradigm and Vision

The IoT represents the concept of bridging the physical world with the virtual world - interconnecting everyday physical objects and allowing them to access Internet services [4]. It is a concept that was initially put forth by Mark Weiser in the early 1990's [5]. Since then, we have seen the advancement of smart devices, network technologies, and a growing push from industry to turn this concept into a reality. But, although close to three decades have passed since this concept was formally discussed, we are still faced with technical and social questions over the IoT. One prominent technical question relates to the constraints of achieving scalable and secure communication between such devices - through existing infrastructures and communication protocols. Nonetheless, this has not stopped the growth in the number of commercially available smart devices such as thermostats or light bulbs, nor deter the drive towards the concepts of smart buildings and smart cities [6–9].

In parallel, we have seen the rapid development and adoption of new mobile technologies which have brought the need for telecommunication networks to support the ever growing number of mobile devices and their continuous demand for data at faster speeds. It is a problem of scalability and throughput which cannot be adequately handled by existing 4G/LTE networks. 5G networks come as a solution to such need. On the whole, it became possible through a combination of new technologies. Five of the most prominent technologies being: (1) Millimeter Waves [10,11], (2) Small Cell Networks [12,13], multi-antenna technologies such as (3) Multiple-Input Multiple-Output (MIMO) [14] and (4) Beamforming [14], and (5) Full Duplex [15,16]. 5G is set to be the foundation for many other emerging technologies, from virtual reality, to autonomous driving, and mission critical application.

In all, the introduction of 5G network technology and the convergence of the IoT paradigm with new network paradigms such as fog computing will bring an unprecedented technology revolution to our world.

3   Real Time Data Utilization Challenges by IoT

IoT devices are gradually entering into different aspects of human life. Personal health monitors as smart-watches [17–19], automated interconnected home appliances, deep-learning driven IoT transformations in medical diagnosis, efficient shopping, defect detection in industries [20] are some of the applications of IoT to name.

According to Yasumoto et al. [21], wireless sensor networks [20] yield multiple IoT data streams that are processed by appropriate software applications. For transmission of IoT data streams various network protocols are used for instance Message Queue Telemetry Transport (MQTT) [22], Constraint Application Protocol (CoAP) [23], Web Socket [24], IPv6 over Low Power Wireless Personal Area Networks (IPv6LoWPAN) [25], MINA [26] etc. Such protocols often give rise to edge-heavy computing which needs extension to fog computing [27] to relieve the pressure of processing and storage near the node devices. The next step is data stream processing. On the basis of [28], the processing is articulated as FIFO queues that respond to data streams and operators i.e. data stream processors handle multiple inputs to and outputs from the FIFO. Different IoT platforms, often supported by machine learning strategies, exist for high-speed real-time processing of vast amounts of temporal data such as IBM Infosphere Stream [29], Amazon AWS IoT, Apache Storm[30] and Spark [31], Microsoft StreamInsight etc.

However, by reference of [21] IoT performance is yet lagging in the fields of heterogeneous data processing, availability of ample IoT platforms with dynamic allocation of resources and granularity customization to mitigate the bandwidth constraint, ability of most platforms to provide automated curation matching and understanding the human version of curation and privacy protection policies applicable for various forms of data. Owing to the problems, the following use-cases cannot be sufficed by current IoT platforms successfully: Firstly, in case of participatory live street view for tourist direction and security services through user, vehicular and street cameras cannot be implemented because cloud resources and network bandwidth are drained out by the incessant upload and downloads of video to the cloud. A new framework allowing parallel flow of multiple data streams to cloud and among stakeholders is necessary instead of convergence. Secondly, ultra-realistic live sports broadcasts through user generated contents is a challenge yet to be resolved due to lack of intelligent and valuable curation of meaningful data from data streams. Thirdly, real-time pedestrian flow tracking in cities for obstructionless vehicular shifting and emergency evacuation is particularly difficult. Not only the latency behind uploading information and processing at the cloud changes the flow of crowd but also the distribution of information to places beyond the location of generation becomes unscalable. Last but not the least, in the monitoring system of the senior citizens, privacy and anomaly detection is of superior concern. Although the activity recognition system by Ueda et al. [32] detects 11 activities with 90% accuracy, the offline learning strategy without the balance of near the edge processing and cloud storage, makes it less effective. Hence 5G should enable the IoT platforms to abstain from the limitations to enhance performance.

4   Machine Type Communications Structure Redesign

Conduloci et al. [33] emphasized on the machine-type multicast service (MtMS) to reconciliate the end-to-end reliability, latency and energy consumption in the up-downlinks. In case of real time applications for instance, smart city, traffic and pollution sensors collect huge amount of data every second. The highly spatial information is later big data processed which in turn leads to highly reliable and fast decisions like changing traffic lights. For such real-time events, connectivity technologies that provide service level agreements (SLAs), i.e., cellular 3GPP technologies [34] such as Long Term Evolution (LTE) [35] and beyond 5G [36] systems are necessary.

The end-to-end delay in MTC environments is caused by uplink(UL), core network and downlink(DL). The UL is utilized to transfer data to remote servers. While the network is in idle state, the data is transmitted through random access (RA) procedure. If two devices send the same preface in the same RA slot, collisions occur. Such collisions and induced delays can be circumvented by access class barring (ACB) [37] and extended access barring (EAB) [38] which introduce backoff mechanism. Although the short access delays to high-priority devices may decline, higher delays incur to other devices. The delays could be reduced through the addition of more preambles rather than only one in the RA as it would abstain from RA reattempts [39].

Management of control and data traffic cause delays in the core network. Although the delay in core network is relatively small, the core network adds delays in the end-to-end communication in the UL and DL directions. Scheduler, frame size, retransmission delay and waiting time for the next transmission frame, round-trip delay in the LTE network act as agents to increment the time by 10–20 ms. The additional delay by the core network in this is as minimal as 1 ms [40]. Activation of UE for UL transfer with or without minimal intervention [41], softwarization, and virtualization in the mobile core of 5G system architecture [42–44, 46] have shown promising performance in reducing overhead and delay, but these approaches need further examination.

In the DL direction of mobile communication, paging procedure [47] is used to alert the MTC devices before data traffic gets delivered. But the capacity of 3GPP to accommodate devices is insufficient. Only 16 devices can receive the page while receiving data. Aiding the problem, there is provision of only two paging occasion in a radio frame of 10 ms. High overhead occurs due to large number of control messages when the number of devices receiving page rises. In order to ameliorate the problem, group paging has been introduced to page a group of devices holding a unique group identity (GID) [38, 48]. But the group paging technique has only been tested coupling it with the legacy 3GPP RA procedure, which in fact, cannot process concurrent multiple entries and in turn increases the collision probability. The back-off methodology for ACB/EAB approaches in UL can be modified and used for DL[37, 49, 50]. However, as in one way the back-off values reduce the chances of collision, in the other way, the delay increases.

The concept of group oriented services led to the Multimedia Broadcast Multicast Services (MBMS) [51] which can endorse group of devices simultaneously for instance mobile TV, video streaming, multimedia content download over mobile networks [52]. The bigger picture is particularly helpful for 5G Architecture to realize the idea of a smart world i.e. smart cities, smart homes, industrial plants, intelligent transportation systems, etc. [53] There are two aspects to be considered to successfully direct the MTC traffic. Firstly, the standard of MBMS is session-oriented. The network provider is responsible for managing a particular MBMS session in precise areas among definite group of devices under a specific client of its own network. Secondly, MBMS is a human-oriented standard. For the formation of multicast groups, human interaction and response is necessary while processing the publication of the MBMS session and joining requests for the session. Additionally, the control traffic has to be redesigned in order to cut the delay of MTC traffic (few bytes) incorporated with the control traffic in the MBMS session.

The design of machine-type multicast service (MtMS) centers the service capability server (SCS) because it handles all the data transmission with MTC device and provides the client with the access to control devices in the group. This decision is particularly advantageous not only in resource utilization in the network but also in reducing delay, overhead and energy consumption in MTC devices. MtMS serving center (MtMS-SC) launches the MtMS session with the list of devices and multicast content at the MtMS gateway (MtMS-GW). The joining request is activated through the mobility management entity (MME) that provides the MtMS coordination entity (MtMS-CE) with the tracking area information of the devices that will receive page with the help of M3 interface. After completion of the joining request, the MtMS-GW finishes the data transfer through the M1 interface. Finally, the values associated with power, modulation and coding scheme are sent to the involved cells through the M2 interface by the MtMS-CE.

The group-paging and code-expanded RA help the idle state to be aware of MtMS traffic in the joining phase. But the utilization of small channel bandwidth aids to the delay and energy consumption as there is limited resource and time allotment in the radio interface. Enhanced group paging is suggested with subgroups of MtMS group to reap the benefits of lower overhead from the group paging. Furthermore, the time interval between two paging message in an RA frame needs to be fixed carefully to keep the overhead delay minimum in 3GPP. Most importantly, 5G will bridge the small bandwidth problem offering a higher frequency range. So MtMS group-paging will hopefully reap the highest profits from the 5G system architecture. The number of paging requests, total delay for paging, average delay for data delivery and data transmission for standard 3GPP paging (SP), group paging (GP) and enhanced group paging (eGP) are considered as the performance metrics here. According to Conduloci et al. [33], firstly, eGP reduces the number of paging requests by 56% than SP for 500 devices (32 requests for SP) and further for larger amount of devices. Secondly, in the total delay for paging, the performance of Standard RA (S-RA) and Code expanded RA (CE-RA) are considered. The limiting aspect of S-RA for all is exhibited in providing multicast service to large number of devices. On the other hand, if broader bandwidths are available for instance, under 5G service GP/CE-RA and Egp/CE-RA will face reduced delays. Thirdly, in case of eGP/CE-RA the average delay in delivery does not escalate with the increase in number of devices and hence proposes the lowest delay, overhead, and energy consumption. Lastly, as more resources are required for data delivery at DL, full coverage is ensured in multicast services until resources for delivery is equivalent to a threshold (25 in reference).

There are some open issues in reduction of data transmission delay that require further focus in future. Firstly, the delay caused in joining procedure to connect the mobile core can be eliminated by migrating the device information to the home-evolved NodeBs gateway through softwarization and virtualization [42–44]. Secondly, an edge-cloud can reduce the end-to-end delay by authorizing the activation of MtMS directly in the edge-cloud. Thirdly, the DL can reduce delay and overhead by analyzing data regarding traffic, network load etc. from the UL. Fourthly, in small channel bandwidth, the appropriate number of UEs to be paged in MtMS should also be figured to reduce the delays. Fifthly, the joining mechanism by a DRX Cycle [45] should be investigated so that all the members of a subgroup in the same DRX Cycle could be paged together. The harmony of UL and DL traffic should be ensured by managing traffic priority. Lastly, parameters like residual battery charge in MTC devices should be taken into account for allocating resources and reduce energy consumption.

5   5G Communication Networks

5.1   Overview

According to the study of Lifeware and Niu et al. [10,54], the 5th generation network utilizes the unique features of the high frequency radio spectrum (in the 30 GHz–300 GHz range) called millimeter waves. Consequently, in 5G communication data transmission will take place at high speed to an enormous extent. Moreover, existing cellular data sparsely occupy the 5G signals. This will enable the signals to be used on account of increasing bandwidth demands in the future. The assistance of small cell to concentrate the network in a smaller range of area and enhance the performance of 5G transmission [12]. Furthermore, massive multiple-input-multiple-output service (mMIMO) will also improve the spectral efficiency through vast increment of antennas and sustain uninterrupted signal reception [12,16]. Large number of antennas in return will attribute in large beamforming returns. The gain will subsidize the collapse of intercell and interstream interference and heighten spectral efficiency.

The 4G towers disperse data in an omnidirectional manner. This phenomenon wastes energy and power in beaming the radio waves at locations where internet access is not required. The 5G towers restrict this occurrence by line-of-sight communications.

5G can accommodate over 1,000 more devices per meter than what was possible by 4G. As 5G signals form shorter wavelengths, the size of antennas has shrunk maintaining the precise directional control. Hence, the same base station will now be capable of supporting more antennas and conclusively, way more devices. Moreover, the nearly doubled channel capacity will enable full duplex technology [15,16]. It contributes in strengthening 5G and contributing to high spectral efficiency.

The end to end latency will be reduced to almost 1 ms in 5G from the around 60 ms latency in 4G. According to EDN Network [55], the 4G LTE cannot bridge this huge gap of latency for three reasons. Firstly, the minimum size for a radio transport block constitutes a sub frame of 1 ms length. This 1 ms duration is entirely spent in the transmission of the block through air interface, without considering processing time at devices and network induced transmission latency. On the contrary, 5G applies a scalable version of orthogonal frequency-division multiplexing with varied numerologies. A 1 ms duration can be accommodated by six different slot configurations of 1, 2, 4, 8, 16 and 32 slots. Hence, if 32 slot configuration is chosen, the minimum size of a radio transport block can be reduced to 0.03125 ms. Secondly, the latency reduction in 4G LTE is obstructed due to the allocation delay of resources between device and base station. The semi-persistent scheduling (SPS) is a unique feature in LTE for periodic data transmissions e.g. voice over IP (VoIP) services. When a device intends to make a data transaction, a resource grant procedure is generated and sent. The time gap between a resource scheduling request and sending data packets is at least 8 ms. However, the request scheduling could be excluded if a mobile handset can utilize the periodic resources while a base station allocates SPS resources.

The device can start transferring data through the preconfigured periodic resource after receiving the data. Thirdly, the LTE SPS setup solely supports a single device. In the event of conditional use of periodic resources e. g. while providing a collision warning by the device, the periodic resources dedicated to the device are wasted as they remain unused during the other timelines. In order to alleviate the aforementioned problems, 5G adopts a special routine based on the LTE SPS service, to share the periodic resources among multiple devise known as a configured grant. The configured grant allocation removes the packet transmission delay incurred due to scheduling request and hence the utilization ratio of allocated periodic resources escalates. As the base station allots multiple users the configured grant, the users or devices randomly avail the resources when they are ready for data transmission.

The benefits of using 5G will touch every sphere of life with a vast impact. 5G can access the currently unreachable places like deep underground, remote, far-away locations. Sensors can be placed there and in times of calamity people can be alarmed. Smart cars, and health monitors will be capable to respond from remote locations to any abnormal situation and alert the user to take necessary decisions with minimum latency [56]. Furthermore, in China, a surgeon performed surgery on an animal by telepresence through 5G network [57]. Most of the concepts that seemed hypothetical and experimental before, will enter into the common world of human lives with the arrival of 5G.

5.2 Challenges and Concerns

Architectural Compromise of 5G and Restoration. In 5G for a successful connection between the antenna and the end device, a clear, line-of-sight propagation is required which complicates the network architecture [10,54]. Beam training is required for transmitting data along the direction of sender and receiver. Moreover, as mmWaves have smaller cells, many of them do not cover great distances as they get absorbed by humidity, rain and blocked by mobile humans even, degrading the network performance. Nonetheless, as 5G uses millimeter waves, the range of data distribution shrinks to almost less than 2% of the range of 4G [70]. Hence, mass installation of 5G towers needs to be implemented in order to ensure reliable transmission of data. Among the other factors, small cell induces self interference and in mMIMO, more antennas produce orthogonal pilot overhead. The overhead exhausts the radio resources. Several approaches to alleviate self interference and architectural framework designs for small cells and MIMO have been proposed in [12–16] to complement 5G.

Impact on Health. 5G uses millimeter waves for data transmission and radio-frequency radiation (RF) is generated as a by product of the use of the wireless technology in phones, wearable and computers. We provide a summary of different views and opinions by scientists and environmentalists in order to articulate how exposure to RF can affect human health. Scientists presume that 5G will cause health issues like abnormal cell division and cell destruction. According to the study [58], RF radiation is carcinogenic [59] and induces tumors [62], cancer [60,61], disrupts gene expressions [63], motor skills, memory and attention [64], interacts with sweat glands [65,66] etc. NYU WIRELESS [67] propose the first temperature-based utilizing magnetic resonance imaging (MRI)-based systems to measure the thermal change. However, the investigation of Guraliuc et al. and Koyama et al. [68,69] perceives an appropriate human body model for dosimetric analysis in 60 GHz band and confirms no significant statistical change in the Human corneal epithelial and human lens epithelial cells in prolonged exposure to RF radiation.

Privacy Concerns. Due to the design of 5G networks, various privacy and security breach issues arise [56]. Firstly, the access point selection theorems at the physical level of 5G, pose a threat towards user location leakage. Violation of location privacy can bring forth semantic information attacks, timing attacks, and boundary attacks. Secondly, International Mobile Subscriber Identity (IMSI) from user equipment can expose the identity of the subscriber. Tracking of Thirdly, 5G integrates various actors for instance, virtual mobile network operators (VMNOs), communication service providers (CSPs), and network infrastructure providers. Each actor has different preference for security and the synchronization among them can be daunting. Fourthly, in 5G, dependence on the new actors e.g. CSP, VMNO etc. leads to sharing the same infrastructure among the actors and security compromise. Hence, the 5G operators will lose governance on the data security and user identity. Fifthly, 5G will liberate the physical boundaries from cloud storage. There will be no guarantee of location of data storage. As different countries enact different privacy policies, data privacy might be violated being stored in the cloud of another country.

Security Countermeasures. The 5G architecture requires superior authentication and reliability levels in order to protect the users from security breach. Firstly, according to Norman et al. [71], IMSI can be secured by the establishment of a pseudonym locally from the user equipment and the home network. The scenario also ensures recovery from lock-out in case a stakeholder has lost the pseudonym.

Lastly, the government should enact the privacy policies according to the need of the country with the collaboration from multinational organizations for instance, the United Nations (UN) and European Union (EU) [72]. Additionally, industry as well as consumer-level regulations require to be ensured, so that groups and individuals can design and enjoy appropriate and necessary levels of privacy and security [72].

6 On Fog Computing

As the network of ubiquitous IoT devices grows and as new smart applications are developed, questions beyond scalable communication between devices and connectivity are brought forth. In particular, we must consider that as the number of devices increases, the network is inundated with huge amount of pragmatic data [73].

The data load is burdensome for traditional storage systems and analytic applications. Hence, the introduction of cloud computing served as a solution that offers scalable processing capacity and on demand storage. However, cloud-only IoT architectures have extensive infrastructure and connectivity limitations. In particular, the process of sending and requesting all the data from the cloud, for either storage or compute, is inadequate for emergency situations. For critical highly-responsive applications, low latency and high throughput of data is a requirement, and a high reliance on the cloud represent a risk. In addition, the scalability will be reduced and network bandwidth faces saturation.

A relative recently proposed solution has been to selectively move computation, storage and control closer to the network edge where data is produced by introducing a system-level architecture. Although, no formal definition has been accepted by the community we can think of the latter as Fog Computing [74]. Fog computing offers a balance between cloud and edge computing - accessing the cloud to eliminate resource contention and utilizing the geographically distributed edge devices when needed. It promotes and balances the programming, communication, analytics, and storage among data centers and end devices. To ensure low latency, it accommodates user mobility, heterogeneous resources and interfaces, and distributed data processing.

According to Bonomi et al. [75] Fog computing is necessary in different fields for instance:

- In cloud service, the implementation details are omitted which is often deemed valuable to reduce latency. However, premium latency applications like gaming, live streaming in virtual/augmented reality, video conferencing etc. require the liberty of accessing the accurate information of where processing or storage occurs if they require.
- Distributed application in different geographical locations as pipeline auditing, environmental change monitor etc.
- Mobile appliances with high speeds as smart cars, connected rails etc.
- Large-scale parallel control systems as smart grid, smart traffic light system etc.

Peng et al. [76] propose a fog computing based radio access network(F-RAN) in 5G network that will relieve the pressure on fronthaul and baseband unit pool and its fast and budget efficient scaling helps F-RAN adjust with the dynamic traffic and radio environment. Tran et al. [77] improves the performance of F-RAN in 5G by formulating a framework that homogenizes the heterogeneous resources at the edge collaborating Mobile Edge Computing (MEC) servers and mobile devices. The advantages of incorporating MEC with Fog computing are demonstrated in the fields of mobile-edge orchestration, collaborative caching and processing, and multi-layer interference cancellation. However, edge caching (already solved and improved by MEC), software-defined networking (SDN), network function virtualization (NFV) [27,76], resource management, interoperability, service discovery, mobility support, fairness, security are some of the open-ended fields for 5G in future studies.

Overall, 5G can be seen as an enabler of fog computing, providing ultra-low latency and high bandwidth to communicate and process data anywhere at the edge of the network. A Fog Computing architecture, in turn, aims provide IoT systems with a smarter and efficient way to interact with compute and storage resources. Furthermore, it can be seen as not just suitable due to technology requirements but also due to business requirements that demand regulatory control over resources. As noted by [78], the use of fog nodes and cloud resources can depend on the domain specific scenario and application. This benefit can be noted when we consider the application of fog computing to enable data processing, e.g AI, closer to data sources such as remote sensors in a smart city. Fog computing could not only reduce the turnaround time of processing data and reduce network costs, which can be heighten in a cloud-based architecture, but also assist in the management of data and devices within a specific jurisdiction.

Currently, the OpenFog Consortium has developed a reference architecture that has been adopted as an official standard by IEEE [78]. This reference, known as IEEE 1934, defines eight core technical pillars: security, scalability, openness, autonomy, RAS (reliability, availability and serviceability), agility, hierarchy and programmability.

7 Enterprise IoT Platforms

Corno et al. [79] have analyzed the services and performances of the current significant IoT platforms based on eight distinguished features- data storage, devices SDK, mobile SDK, push notifications, Rest APIs, supported protocols, virtual devices, and analytics. The comparison also provides an overview of what services the IoT platforms can deliver and their applications.

Xively, now empowered by google, employs a unique IoT platform as a service [80]. It provides template application for mobile phones [79]. The platform does not provide data storage at its end and the push notication for generating alerts is minimum [79]. Xively uses the MQTT, HTTP protocols and MQTT provides the virtual support for device twin [79].

Bosch and Arrayent IoT platforms provide limited functionality [79]. Bosch lacks the data storage, and Mobile SDK support. On the other hand, Arrayent IoT cloud services do not support virtual devices, hardware SDK and storage of data at the platform. For Bosch IoT suite, HTTP, MQTT, LWM2M, mPRM network protocols are followed; IoT remote manager handles the hardware SDK; Java client or HTTP API Manager integrate with software applications; IoT developer console performs data analysis; push notifications are generated for remote events [79]. Arrayent abides by HTTPS, and web sockets as protocol; is accessed by mobile phones through iOS and android; its REST API follows EcoAdaptor framework; covers data analytics service; generates realtime alerts [79].

AWS IoT Core, Google Cloud IoT, IBM Watson IoT, Microsoft Azure IoT, Oracle IoT Cloud Service, and SAP IoT platforms offers different generic cloud services along with IoT ecosystem [79]. The push notification is mostly generated by MQTT subscription [79]. They offer mostly the same set of sevices [79].

Google Cloud IoT platform follows the MQTT, and HTTP protocols [79]. It accommodates both hardware and mobile SDK with the REST support from Google Cloud IoT API. Cloud Pub/Sub (7 days) implement the device twins [79]. Google has its own cloud and Data Studio for big data analysis [79]. Lastly, Cloud Pub/Sub generates push notifications [79].

SAP IoT platform does not follow specific protocols [79]. It provides hardware support, Mobile SDK is enterprised by cloud platform and iOS extended by REST API and storage at the platform [79]. It uses the Apple Push Notification services for alerts and messages [79]. SAP utilizes the Thing Registry for virtual device [79].

General Electrics has a more focused set of functions [79]. It follows the MQTT, HTTP protocol [79] for network. For Hardware and Mobile SDK support, there is provision for hybrid Predix Machine [79]. The REST API is bolstered by the asset services [79]. Mobile gateway acts as the device twin and data is stored at Blobstore [79]. However, data analysis and push notification services are missing [79].

IBM Watson IoT follows the same MQTT and HTTP as the General Electrics and the Google Cloud IoT platform [79]. Data analysis is performed by MQTT Watson IoT and data is stored at the Bluemix storage [79]. Push Notification is developed by Bluemix Push Notifications [79]. It supports Edge analytics SDK in Hardware respect and Android applications for mobile. Virtual devices are implemented through MQTT [79].

The thinger.io IoT platform abides by the HTTPs protocol [79] and as hardware SDK, can access Arduino, Sigfox or Linux SDK and Android applications alongwith Server API for mobile [79]. It can access device twin virtually [79] and contains data bucket [79]. Data analytics is performed by cloud console [79] and has no push-notification service [79].

Micosoft Azure IoT pursues AMQP along with the popular MQTT and HTTPS [79]. The device provisioning strategy is implied at the hardware sector and both Android and iOS applications are supported [79]. Besides, data analysis, it also stores data at the Azure storage [79]. It uses IoT edge to develop device twin and alerts the cloud through Notification hub [79].

Oracle IoT Cloud Service supports device twinning [79], stores data at the platform and allows notification to the cloud [79]. It follows MQTT and HTTPs as protocols [79]. It constitutes Endpoint management to handle hardware and uses Java and iOS to access mobile phones [79].

AWS IoT core anticipated as one of the pioneering data clouds and IoT platform, is highly conducive for developing industries [86,87]. It follows HTTP, MQTT and web-socket for network protocols. It can access both android and iOS applications and connects hardware via AWS Greengrass [79]. The data is stored in S3 and analysed at AWS Console. For any event, notifications are sent to the cloud through Amazon SNS.

Last but not the least, for fast and reactive smart applications, the reputation of Cisco is rising day by day [82,85]. It is the only platform that supports Fog computing till date [85]. It supports independent hardware vendors [82] and can allow access to iOS applications through programmable APIs [83]. It utilizes gateway to reinforce virtual device [84]. Besides, data storage and analysis operations, Cisco IoT platform sends notification to cloud on instance of an event [82,83,85].

8   Implication of 5G on IoT Platforms

In the following table we investigate and compare the aforementioned IoT Platforms on the basis of their fundamental features and a 5G component- Fog Computing.

Table 1. Relative study of the features of popular IoT platforms

Platforms	Protocol	Hardware Support	Mobile Services	REST API	Virtual Device	Storage At the Platform	Analytics	Push Notifications	Fog Computing
Xively [78, 79]	MQTT, HTTP	Present	Readymade Template applications	Present	MQTT	Absent	Present	Minimal Push Notification for Alert	Absent
Bosch IoT Suite [78]	HTTP, MQTT, LWM2M, mPRM	IoT Remote Manager	Absent	Java client or HTTP API Manager	Present	Absent	IoT Developer Console	Remote Event Push	Absent
Google Cloud IoT Platform [78]	MQTT, HTTP	Present	Present	Google Cloud IoT API	Cloud Pub/ Sub (7 days)	Cloud Storage	Google Data Studio	Cloud Pub/ Sub	Absent
SAP IoT Platform [78]	Absent	Present	Cloud Platform, iOS	Present	Thing Registry	Present	Absent	Apple Push Notification Service	Absent
Arrayent IoT Cloud Services [78]	HTTPS, Web sockets	Absent	Android, iOS	EcoAdaptor framework	Absent	Absent	Present	Realtime Alerts	Absent
General Electrics Predix [78]	HTTPS, Web Sockets	Predix Machine	Predix SDK for Hybrid device	Asset Services	Mobile gateway	Blobstore (S3)	Absent	Absent	Absent
IBM Watson IoT [78]	MQTT, HTTP	Edge Analytics SDK	Android	Present	MQTT	Bluemix Storage	MQTT Watson IoT	Bluemix Push Notifications	Absent
thinger.io [78]	HTTPs	Arduino, Sigfox Or Linux	Android application	Server API	Present	Data Bucket	Cloud Console	Absent	Absent
Microsoft Azure IoT [78]	MQTT, HTTPS, AMQP	Device Provisioning	Android, iOS	Present	IoT Edge	Azure Storage	Present	Notification Hubs	Absent
Oracle IoT Cloud Service [78]	MQTT, HTTPs	Endpoint Management	Java, iOS	Present	Present	Present	Present	Present	Absent
Amazon or AWS IoT core [78]	HTTP, MQTT, Web Socket	AWS Greengrass	Android, iOS	Present	Present	S3	AWS Console	Amazon SNS	Absent
Cisco [80-84]	Present	Independent Hardware Vendors	iOS	Open and Programmable APIs	Gateway	Present	Present	Event Response	Present

From Table 1, the lack of implementation of Fog computing in all the platforms except Cisco is easily distinguishable. Currently, Microsoft Azure, and AWS IoT are the most popular platforms for cutting edge innovative applications in industries [86,87]. But in the event of highly responsive systems making decisions within seconds, Fog computing takes the lead and is clearly the merger of 4G and 5G. Let us consider smart transportation systems. Commercial Jet Planes generate 10 TB of data per 30 min [85]. If the data were stored in cloud, analyzed and necessary data had to be retrieved, the whole process could occupy a time span of minutes to days. When response is time sensitive (less than a second) and mass data is collected at the edge, Fog computing is the best solution to secure and maintain communication effectively. Hence, Cisco IoT platform is largely accepted to implement smart cities and smart vehicular transportation systems due to its sensitive reactivity and conformity with 5G communication [82, 85].

9   Discussion

The IoT will demand a computing architecture that is more decentralized than current computing and data storage models.

Among different platforms, Cisco is the pioneer in integrating Fog computing into its feature set. Fog computing balances the computational procedures and storage among the edge and cloud. It is an instinctive characteristic of 5G which will diminish the latency and result in fast communication. In sensitive and reactive systems for instance, commercial jet planes [85] about 10TB data gets transmitted every half an hour. Propagating to and from cloud the cloud disrupts the fast decision-making and increases latency. Hence smart cars and other intelligent vehicles are using Fog computing. The purpose of the survey is to use a suitable IoT platform for a remotely controlled telepresence robots. The robots will access disastrous environment, rescue the victims and react to critical situations. We shall build and test our robots implementing Cisco IoT platform to fulfill our requirements and be responsive to the surroundings being remotely controlled.

When looking into the future trends of compute services, we must look into Amazon Web Services, Google Cloud, Microsoft Azure and Cisco's Cloud Service. Inherently, these four tech giants are also leading the pack in IoT services. From a business perspective, the increasing demands for storage and compute from these devices presents a profitable opportunity. Nonetheless, these tech giants must also transform their technology and business models as the IoT begins to demand for a more decentralized infrastructure that provides higher quality of service. As discussed, fog computing is one technology trend that promises to significantly reduce the amount of processing power required from the cloud and shift it to the edge of the network.

As the leader in cloud computing [88, 89], as of this writing, Amazon's IoT services offer far greater number of developer tools and cloud computing features than its rivals, and has a dedicated IoT edge computing service called AWS IoT Edge [90, 91]. With the AWS IoT platform, devices can communicate with application running in AWS instances like Lambda, and use communication protocol like HTTP, MQTT or WebSockets. As part of their IoT Platform, Amazon offers IoT Greengrass [91, 92], which extends AWS to edge devices allowing them to act locally on the data they generate, while still using the cloud for management, analytics, and persistent storage. This, in essence, means that devices connected to Greengrass can respond quickly to local events, interact with local resources and connect to the cloud when needed, e.g. for backing up storage, or extended analytics. Other services include AWS IoT Core which enables simple and secure interaction between devices [93]. Additionally, Amazon provides Amazon FreeR-TOS - a microcontroller operating system for low-power edge devices that allows for seamless integration with other Amazon IoT services such as IoT Core and Greengrass.

Similar to Amazon, Google has it's own IoT platform - Google Cloud IoT Platform [94–97]. Google Cloud IoT Core serves as the backbone to Google's IoT service offerings, allowing developer to securely connect, manage and process data from millions of globally connected devices. Similar to Amazon's Lambda Functions, Cloud IoT Core run's on Google's serverless infrastructure, providing an infrastructure that scales and responds to real-tie events. Currently, Cloud IoT Core only support MQTT and HTTP. Similar to Greengrass, Google's Cloud IoT Edge service allows edge devices to respond to real-time events and make decisions without continuously communicating to the cloud. The latter two IoT services can be connected to other Google Cloud services such as Cloud Pub/Sub or Firebase.

A key differentiator for Microsoft Azure IoT service's is its Azure IoT Protocol Gateway which adapts incoming and outgoing traffic to comply to a particular devices communication protocol (e.g. AMPQ, MQTT and HTTP).

The target of the analysis of different platforms further empowered by 5G is to implement one of them in building a telepresence robots. Based on sensitive response, power efficiency, data management, and access to Fog computing, Cisco has been proposed to assuage the purpose [81–85]. As mentioned earlier, only Cisco enacted Fog computing into their platform. Fog computing authorizes a balance in the data storage, assembly, analysis and interpretation between the edge and cloud. The intelligent data processing technique offered by Fog computing will support the telepresence robots to react spontaneously and make decisions to handle dangerous situations swiftly. Moreover, the speed and management of huge volumes of data by 5G network will contribute further to realize the goal of a realtime autonomous as well as instinctive robot adept to dealing with emergencies.

10   Conclusion

The IoT represents the vision of an interconnected world of devices and humans. The development of fifth-generation communication networks brings us a step closer to such vision by enabling faster and more reliable connectivity. In parallel, it further enables other technological advancement that are necessary to bring forth an interconnected world. It further emancipates the IoT to be preached at the broadened spheres of technological worlds. The scrutinizing of various platforms, provides with an insight of their features and capabilities for different goals. Therefore, Cisco aims at fulfilling the motivation of an environment sensitive and fast-reacting telepresence robots for emergencies. The analytical survey shall provide researchers to gather knowledge on distinct platforms and aid in realizing their projects using the appropriate.

References

1. Featured page of Internet of Things Wiki. https://internetofthingswiki.com/top-20-iot-platforms/634/. Accessed 20 Apr 2019
2. OnQ Blog of Qualcomm. http://tiny.cc/wnyj5y. Accessed 20 Apr 2019
3. Updated List IoT Platforms in 2017 by IoT Analytics. https://iot-analytics.com/iot-platforms-company-list-2017-update/. Accessed 21 Apr 2019
4. Mattern, F., Floerkemeier, C.: From the internet of computers to the Internet of Things. In: Sachs, K., Petrov, I., Guerrero, P. (eds.) From Active Data Management to Event-Based Systems and More. LNCS, vol. 6462, pp. 242–259. Springer, Heidelberg (2010). https://doi.org/10.1007/978-3-642-17226-7_15
5. Weiser, M.: The computer for the 21st century. IEEE Pervasive Comput. $1(1)$, 19–25 (2002). https://doi.org/10.1109/MPRV.2002.993141
6. Pan, J., Jain, R., Paul, S., Vu, T., Saifullah, A., Sha, M.: An internet of things framework for smart energy in buildings: designs, prototype, and experiments. IEEE Internet Things J. $2(6)$, 527–537 (2015). https://doi.org/10.1109/JIOT.2015.2413397
7. Hu, X., Yang, L., Xiong, W.: A novel wireless sensor network frame for urban transportation. IEEE Internet Things J. $2(6)$, 586–595 (2015). https://doi.org/10.1109/JIOT.2015.2475639
8. Handte, M., Foell, S., Wagner, S., Kortuem, G., Marrón, P.J.: An internet-of-things enabled connected navigation system for urban bus riders. IEEE Internet of Things J. $3(5)$, 735–744 (2016). https://doi.org/10.1109/JIOT.2016.2554146
9. Minoli, D., Sohraby, K., Occhiogrosso, B.: IoT considerations, requirements, and architectures for smart buildings–energy optimization and next-generation building management systems. IEEE Internet Things J. $4(1)$, 269–283 (2017). https://doi.org/10.1109/JIOT.2016.2554146
10. Niu, Y., Li, Y., Jin, D., Su, L., Vasilakos, A.V.: A survey of millimeter wave communications (mmWave) for 5G: opportunities and challenges. Wireless Netw. $21(8)$, 2657–2676 (2015). https://doi.org/10.1007/s11276-015-0942-z
11. Sun, S., Rappaport, T.S., Shafi, M., Tang, P., Zhang, J., Smith, P.J.: Propagation models and performance evaluation for 5G millimeter-wave bands. IEEE Trans. Veh. Technol. $67(9)$, 8422–8439 (2018). https://doi.org/10.1109/TVT.2018.2848208
12. Jungnickel, V., et al.: The role of small cells, coordinated multipoint, and massive MIMO in 5G. IEEE Commun. Mag. $52(5)$, 44–51 (2014). https://doi.org/10.1109/MCOM.2014.6815892
13. Zhang, H., Dong, Y., Cheng, J., Hossain, M.J., Leung, V.C.: Fronthauling for 5G LTE-U ultra dense cloud small cell networks. IEEE Wirel. Commun. $23(6)$, 48–53 (2016). https://doi.org/10.1109/MWC.2016.1600066WC
14. Vook, F.W., Ghosh, A., and Thomas, T.A.: MIMO and beamforming solutions for 5G technology. In: 2014 IEEE MTT-S International Microwave Symposium (IMS 2014), pp. 1–4. IEEE (2014). https://doi.org/10.1109/MWSYM.2014.6848613

15. Zhang, Z., Chai, X., Long, K., Vasilakos, A.V., Hanzo, L.: Full duplex techniques for 5G networks: self-interference cancellation, protocol design, and relay selection. IEEE Commun. Mag. **53**(5), 128–137 (2015). https://doi.org/10.1109/MCOM. 2015.7105651

16. Xia, X., Xu, K., Wang, Y., Xu, Y.: A 5G-enabling technology: benefits, feasibility, and limitations of in-band full-duplex mMIMO. IEEE Veh. Technol. Mag. **13**(3), 81–90 (2018). https://doi.org/10.1109/MVT.2018.2792198

17. Health options in Apple Watch Series 4. https://www.apple.com/apple-watch-series-4/health/. Accessed 27 Apr 2019

18. Health features in fitbit. https://www.wired.com/story/when-your-activity-tracker-becomes-a-personal-medical-device/. Accessed 27 Apr 2019

19. Health Monitoring System in Samsung Gear S3 Frontier. https://support.t-mobile. com/docs/DOC-33560. Accessed 27 Apr 2019

20. Transformation IoT features page. http://tiny.cc/gyou5y. Accessed 28 Apr 2019

21. Yasumoto, K., Yamaguchi, H., Shigeno, H.: Survey of real-time processing technologies of IoT data streams. J. Inf. Process. **24**(2), 195–202 (2016). https://doi. org/10.2197/ipsjjip.24.195

22. COASIS Standard, MQTT version 3.1.1 (2014). http://docs.oasis-open.org/mqtt/ mqtt/v3.1.1/os/mqtt-v3.1.1-os.doc

23. Shelby, Z., Hartke, K., Bormann, C.: Request for Comment 7252, The Constrained Application Protocol (CoAP) (2014). http://tools.ietf.org/rfc/rfc7252.txt

24. WebSocket. https://www.websocket.org/. Accessed 28 Apr 2019

25. Shelby, Z., Borman, C.: 6LoWPAN: The Wireless Embedded Internet. Wiley, Chichester (2011)

26. Qin, Z., Denker, G., Giannelli, C., Bellavista, P., Venkatasubramanian, N.: A software defined networking architecture for the Internet-of-Things. In: 14th Proceedings IEEE Network Operations and Management Symposium (NOMS), pp. 1–9 (2014). https://doi.org/10.1109/NOMS.2014.6838365

27. Luan, T.H., Gao, L., Li, Z., Xiang, Y., Wei, G., Sun, L.: Fog computing: focusing on mobile users at the edge. arXiv preprint, arXiv:1502.01815, 6 February 2015

28. Hirzel, M., Soulé, R., Schneider, S., Gedik, B., Grimm, R.: A catalog of stream processing optimizations. ACM Comput. Surv. **46**(4), 1–34 (2014). https://doi. org/10.1145/2528412. Article 46

29. StreamBase Systems. http://www.streambase.com (2012). Accessed 28 Apr 2019

30. Storm project. http://storm-project.net/ (2012). Accessed 28 Apr 2019

31. Documentation of Apache Spark-Lighting-fast cluster computing. http://spark. apache.org/

32. Ueda, K., Suwa, H., Arakawa, Y., Yasumoto, K.: Exploring accuracy-cost trade-off in in-home living activity recognition based on power consumptions and user positions. In: 14th Proceedings of IEEE International Conference on Ubiquitous Computing and Communications (IUCC 2015), pp. 1131–1137 (2015). https://doi. org/10.1109/CIT/IUCC/DASC/PICOM.2015.169

33. Condoluci, M., Araniti, G., Mahmoodi, T., Dohler, M.: Enabling the IoT machine age with 5G: machine-type multicast services for innovative real-time applications. IEEE Access **4**, 5555–5569 (2016). https://doi.org/10.1109/ACCESS.2016.2573678

34. Palattella, M.R., et al.: Internet of Things in the 5G era: enablers, architecture, and business models. IEEE J. Sel. Areas Commun. **34**, 510–527 (2016). https:// doi.org/10.1109/JSAC.2016.2525418

35. 3GPP: Evolved Universal Terrestrial Radio Access (E-UTRA) and Evolved Universal Terrestrial Radio Access Network (E-UTRAN); Overall description, TS 36.300

36. Boccardi, F., Heath, R.W., Lozano, A., Marzetta, T.L., Popovski, P.: Five disruptive technology directions for 5G. IEEE Commun. Mag. **52**, 74–80 (2014). https://doi.org/10.1109/MCOM.2014.6736746
37. 3GPP: Evolved Universal Terrestrial Radio Access (E-UTRA); Radio Resource Control (RRC), TS 36.331
38. 3GPP: RAN improvements for machine-type communications, TR 37.868
39. Thomsen, H., Pratas, N.K., Stefanovic, C., Popovski, P.: Code-expanded radio access protocol for machine-to-machine communications. Trans. Emerging Telecommun. Technol. **24**(4), 355–365 (2013). https://doi.org/10.1002/ett.2656
40. Blajić, T., Nogulić, D., and Družijanić, M.: Latency Improvements in 3G Long Term Evolution, in Mipro CTI, svibanj (2006)
41. 3GPP: Study on Enhancements to Machine-Type Communications (MTC) and Other Mobile Data Applications; Radio Access Network (RAN) Aspects, TR 37.869
42. Li, Y., Chen, M.: Software-defined network function virtualization: a survey. IEEE Access **3**, 2542–2553 (2015). https://doi.org/10.1109/ACCESS.2015.2499271
43. Wood, T., Ramakrishnan, K.K., Hwang, J., Liu, G., Zhang, W.: Toward a software-based network: integrating software defined networking and network function virtualization. IEEE Network **29**, 36–41 (2015). https://doi.org/10.1109/MNET.2015.7113223
44. Mijumbi, R., Serrat, J., Gorricho, J.L., Bouten, N., Turck, F.D., Boutaba, R.: Network function virtualization: state-of-the-art and research challenges. IEEE Commun. Surv. Tutorials **18**, 236–262 (2016). https://doi.org/10.1109/COMST.2015.2477041. Firstquarter
45. Bontu, C.S., Illidge, E.: DRX mechanism for power saving in LTE. IEEE Commun. Mag. **47**(6), 48–55 (2009). https://doi.org/10.1109/MCOM.2009.5116800
46. Aissioui, A., Ksentini, A., Gueroui, A.M., Taleb, T.: Toward elastic distributed SDN/NFV controller for 5G mobile cloud management systems. IEEE Access **3**, 2055–2064 (2015). https://doi.org/10.1109/ACCESS.2015.2489930
47. 3GPP: Mobile radio interface layer 3 specification, core network protocols; Stage 2, TS 23.108
48. Wei, C.H., Cheng, R.G., Tsao, S.L.: Performance analysis of group paging for machine-type communications in LTE networks. IEEE Trans. Veh. Technol. **62**, 3371–3382 (2013). https://doi.org/10.1109/TVT.2013.2251832
49. Arouk, O., Ksentini, A., Taleb, T.: Group paging optimization for machine-type-communications. In: IEEE International Conference on Communications (ICC), pp. 6500–6505, June 2015. https://doi.org/10.1109/ICC.2015.7249360
50. Arouk, O., Ksentini, A., Taleb, T.: Group paging-based energy saving for massive MTC accesses in LTE and Beyond networks. IEEE J. Sel. Areas Commun. **PP**(99), 1–1 (2016). https://doi.org/10.1109/jsac.2016.2520222
51. Lecompte, D., Gabin, F.: Evolved multimedia broadcast/multicast service (eMBMS) in LTE-advanced: overview and Rel-11 enhancements. IEEE Commun. Mag. **50**, 68–74 (2012). https://doi.org/10.1109/MCOM.2012.6353684
52. Condoluci, M., Araniti, G., Molinaro, A., Iera, A.: Multicast resource allocation enhanced by channel state feedbacks for multiple scalable video coding streams in LTE networks. IEEE Trans. Veh. Technol. **PP**(99), 1–1 (2015). https://doi.org/10.1109/TVT.2015.2449080
53. Perera, C., Liu, C.H., Jayawardena, S., Chen, M.: A survey on Internet of Things from industrial market perspective. IEEE Access **2**, 1660–1679 (2014). https://doi.org/10.1109/ACCESS.2015.2389854
54. Difference between 4G and 5G. https://www.lifewire.com/5g-vs-4g-4156322. Accessed 20 Apr 2019

55. Study of reduced latecy of 5G waves by EDN Network. http://tiny.cc/rkwm5y. Accessed 17 Apr 2019

56. Ahmad, I., Kumar, T., Liyanage, M., Okwuibe, J., Ylianttila, M., Gurtov, A.: Overview of 5G security challenges and solutions. IEEE Commun. Stand. Mag. **2**(1), 36–43 (2018)

57. News on Remote Robotic Surgery in China. http://tiny.cc/h4mt5y. Accessed 25 Apr 2019

58. 5G And The IOT: Scientific Overview Of Human Health Risks. https://ehtrust. org/key-issues/cell-phoneswireless/5g-networks-iot-scientific-overview-human-health-risks/. Accessed 30 Apr 2019

59. Miller, A.B., Morgan, L.L., Udasin, I., Davis, D.L.: Cancer epidemiology update, following the 2011 IARC evaluation of radiofrequency electromagnetic fields (Monograph 102). Environ. Res. **167**, 673–683 (2018). https://doi.org/10.1016/ j.envres.2018.06.043

60. Hardell, L., Carlberg, M.: Comments on the US National toxicology program technical reports on toxicology and carcinogenesis study in rats exposed to whole-body radiofrequency radiation at 900 MHz and in mice exposed to whole-body radiofrequency radiation at 1,900 MHz. Int. J. Oncol. **54**(1), 111–127 (2019). https://doi. org/10.3892/ijo.2018.4606

61. Peleg, M., Nativ, O., Richter, E.D.: Radio frequency radiation-related cancer: assessing causation in the occupational/military setting. Environ. Res. **163**, 123–133 (2018). https://doi.org/10.1016/j.envres.2018.01.003

62. Lerchl, A., et al.: Tumor promotion by exposure to radiofrequency electromagnetic fields below exposure limits for humans. Biochem. Biophys. Res. Commun. **459**(4), 585–590 (2015). https://doi.org/10.1016/j.bbrc.2015.02.151

63. Di Ciaula, A.: Towards 5G communication systems: are there health implications? Int. J. Hyg. Environ. Health **221**(3), 367–375 (2018). https://doi.org/10.1016/j. ijheh.2018.01.011

64. Meo, S.A., Almahmoud, M., Alsultan, Q., Alotaibi, N., Alnajashi, I., Hajjar, W.M.: Mobile phone base station tower settings adjacent to school buildings: impact on students' cognitive health. Am. J. Men's Health **13**(1), 1557988318816914 (2019). https://doi.org/10.1177/1557988318816914

65. Betzalel, N., Feldman, Y., Ishai, P.B.: The modeling of the absorbance of sub-THz radiation by human skin. IEEE Trans. Terahertz Sci. Technol. **7**(5), 521–528 (2017). https://doi.org/10.1109/TTHZ.2017.2736345

66. Betzalel, N., Ishai, P.B., Feldman, Y.: The human skin as a sub-THz receiver-Does 5G pose a danger to it or not? Environ. Res. **163**, 208–216 (2018). https://doi. org/10.1016/j.envres.2018.01.032

67. NYU Research mmWave Health Effects. https://wireless.engineering.nyu.edu/ mmwave-health-effects/. Accessed 29 Apr 2019

68. Guraliuc, A.R., Zhadobov, M., Sauleau, R., Marnat, L., Dussopt, L.: Millimeter-wave electromagnetic field exposure from mobile terminals, In: 2015 European Conference on Networks and Communications (EuCNC), pp. 82–85. IEEE (2015). https://doi.org/10.1109/EuCNC.2015.7194045

69. Koyama, S., et al.: Effects of long-term exposure to 60 GHz millimeter-wavelength radiation on the genotoxicity and heat shock protein (Hsp) expression of cells derived from human eye. Int. J. Environ. Res. Public Health **13**(8), 802 (2016). https://doi.org/10.3390/ijerph13080802

70. Webpage on Challenges of 5G. http://tiny.cc/48sz5y. Accessed 28 Apr 2019

71. Norrman, K., Näslund, M., Dubrova, E.: Protecting IMSI and user privacy in 5G networks. In: 9th Proceedings of EAI International Conference on Mobile Multimedia Communications Institute for Computer Science, Social-Informatics, and Telecommunication Engineering, pp. 159–66 (2016). https://doi.org/10.4108/eai. 18-6-2016.2264114

72. Ahmad, I., Kumar, T., Liyanage, M., Okwuibe, J., Ylianttila, M., Gurtov, A.: 5G security: analysis of threats and solutions. In: IEEE Conference Standards for Communication and Networking, pp. 193–99, September 2017. https://doi.org/10. 1109/CSCN.2017.8088621

73. Dastjerdi, A.V., Buyya, R.: Fog computing: helping the Internet of Things realize its potential. Computer **49**(8), 112–116 (2016). https://doi.org/10.1109/MC.2016. 245

74. Shi, W., Dustdar, S.: The promise of edge computing. Computer **49**(5), 78–81 (2016). https://doi.org/10.1109/MC.2016.145

75. Bonomi, F., Milito, R., Natarajan, P., Zhu, J.: Fog computing: a platform for Internet of Things and analytics. In: Bessis, N., Dobre, C. (eds.) Big Data and Internet of Things: A Roadmap for Smart Environments. SCI, vol. 546, pp. 169–186. Springer, Cham (2014). https://doi.org/10.1007/978-3-319-05029-4_7

76. Peng, M., Yan, S., Zhang, K., Wang, C.: Fog computing based radio access networks: issues and challenges, 13 June 2015. https://doi.org/10.1109/MNET.2016. 7513863

77. Tran, T.X., Hajisami, A., Pandey, P., Pompili, D.: Collaborative mobile edge computing in 5G networks: new paradigms, scenarios, and challenges, 9 December 2016. https://doi.org/10.1109/MCOM.2017.1600863

78. OpenFog Consortium Architecture Working Group, https://www. openfogconsortium.org/wp-content/uploads/OpenFog-Architecture-Overview-WP-2-2016.pdf. Last Accessed on 29 Apr 2019

79. Corno, F., De Russis, L., Sáenz, J.P.: On the advanced services that 5G may provide To IoT applications. In: 2018 IEEE 5G World Forum (5GWF), pp. 528–531, July 2018. https://doi.org/10.1109/5GWF.2018.8517038

80. Ray, P.P.: A survey of IoT cloud platforms. Fut. Comput. Inf. J. (2017). https:// doi.org/10.1016/j.fcij.2017.02.001

81. Cisco Kinetic for Manufacturing Harnessing IoT data to boost productivity. http:// tiny.cc/nv3y5y. Accessed 29 Apr 2019

82. Smart City framework by Cisco. http://tiny.cc/bh3y5y. Accessed 29 Apr 2019

83. Architecture of Cisco IoT Platform. http://tiny.cc/2e3y5y. Accessed 29 Apr 2019

84. Homepage of Cisco Kinetic. http://tiny.cc/092y5y. Accessed 29 Apr 2019

85. Fog Computing and the Internet of Things: Extend the Cloud to Where the Things Are (2015). http://tiny.cc/nv3y5y. Accessed 29 Apr 2019

86. Comparison between Microsoft Azure and Amazon Web Services. https://stackify. com/azure-vs-aws-comparison/. Accessed 29 Apr 2019

87. Overview of Microsoft Azure and Amazon Web Services. https://azure.microsoft. com/en-us/overview/azure-vs-aws/. Accessed 29 Apr 2019

88. Cusumano, M.A.: Cloud computing and SaaS as new computing platforms. Commun. ACM **53**(4), 27–29 (2010). https://doi.org/10.1145/1721654.1721667

89. Smith, R.: Computing in the cloud. Res. Technol. Manag. **52**(5), 65–68 (2009). https://doi.org/10.1080/08956308.2009.11657590

90. Amazon Makes Foray Into Edge Computing With AWS Greengrass. http://tiny. cc/ft1y6y. Accessed 10 May 2019

91. AWS IoT Greengrass. https://aws.amazon.com/greengrass/. Accessed 11 May 2019

92. What Is AWS IoT Greengrass? https://docs.aws.amazon.com/greengrass/latest/developerguide/what-is-gg.html. Accessed 11 May 2019
93. AWS IoT Core Documentation. https://docs.aws.amazon.com/iot/index.html. Accessed 11 May 2019
94. CLOUD IOT CORE. http://tiny.cc/3e3y6y. Accessed 15 May 2019
95. Google Cloud IoT. https://cloud.google.com/solutions/iot/. Accessed 15 May 2019
96. IoT in the Google Cloud. https://www.qwiklabs.com/quests/49. Accessed 15 May 2019
97. Cloud IoT Core overview. https://cloud.google.com/iot/docs/concepts/overview. Accessed 15 May 2019

Optical Wireless Transmission
of Electrocardiogram During Effort

Stéphanie Sahuguède[1]([⊠]), Anne Julien-Vergonjanne[1], Olivier Bernard[2],
Kostiantyn Vasko[3], and Boris Shtangei[3]

[1] University of Limoges, CNRS, XLIM, UMR 7252, 87000 Limoges, France
stephanie.sahuguede@unilim.fr
[2] University of Poitiers, MOVE, 86000 Poitiers, France
[3] Kharkiv National University of Radio Electronics, Nauky Ave. 14, Kharkiv 61166, Ukraine

Abstract. In the context of contactless diagnostics of human bio-parameters,
wireless transmission are generally based on radio-frequency technologies. This
can be a problem regarding interferences with other equipments or with interactions for human body as for example people wearing pacemakers. We present in
this paper the realization of a portable device able to perform distant electrocardiogram (ECG) monitoring during physical activities based on optical wireless communications. Our system is designed using on-the-shelf components and imposes
constraints on sampling frequency. Therefore, even if the sampling frequency in
the proposed system is not sufficient for medical diagnosis, our study shows the
potentiality of using optical wireless for transmission and distant analysis of ECG
data. We illustrate the performance by experimentally evaluating packet error rate
of the optical wireless transmission of ECG data and we show that the heart rate
is correctly evaluated when the person is moving inside a room.

Keywords: Optical wireless communication · Wireless ECG transmission ·
Packet Error Rate

1 Introduction

Nowadays, there is a rapid development of medical technology. One of the most developing branches is contactless diagnostics of human bio parameters. With the help of technology, we can track many of the characteristics of a human condition, for example - the
bio potentials of the heart, temperature, heart rate and physical activity. Electrocardiography is one of the oldest and most common methods of studying the heart. Performing the
electrocardiogram requires specific equipment. In hospitals, electrocardiogram (ECG)
is generally still interpreted using a recorder, which has several drawbacks: it is difficult
to conduct a comparative analysis of the electrocardiogram at different periods, it is necessary to use special chart paper, there is no automated diagnosis and storage of patient
data. Existing devices able to record ECG data are generally based on wired solutions
and based on several leads using more than 3 electrodes for medical diagnosis. This type
of device is not adapted to be used during daily life activity with body motion. Therefore,
there is a need to develop inexpensive and accurate portable electrocardiograph.

© Springer Nature Switzerland AG 2020
O. Habachi et al. (Eds.): UNet 2019, LNCS 12293, pp. 226–233, 2020.
https://doi.org/10.1007/978-3-030-58008-7_18

In this context, in order to enhance comfort of use and reduce disturbance due to cables, the development of wireless devices for monitoring and recording human physical activity is relevant, not only to be able to record history of the data numerically but also to represent and analyze data during efforts. In order to wirelessly transmit ECG data, several challenges have to be taken into account: frequency sampling (so transmission rate) sufficiently high in order to ensure a good ECG quality [1, 2] and miniaturization of circuits for emission and reception in order to integrate them near from the electrodes. Typically, a sampling frequency of minimal 250 Hz is required for medical purpose since most of the diagnostic information is contained below 100 Hz for adults [1]. For heart rate evaluation, the sampling frequency can be much lower since the fundamental frequency for the QRS complex is around 10 Hz [1].

Several works have been done on wireless monitoring of ECG based on radio frequency (RF) technologies [2–6]. RF communications (Bluetooth, zigbee, wifi, ...) reliability is not always ensured especially due to interference with other technologies using same band frequencies. In addition, radio wave impact on health is controversial but unknown, so it is interesting to investigate alternative solutions such as optical wireless communications (OWC). Provided that the constraints about eye safety are satisfied, OWC are safe and not subjected to RF interference. Using OWC technology has gain much interest last years and some work have proposed its application for monitoring health related parameter [7–13]

Published works about optical wireless ECG devices are all dealing with visible light communication (VLC) [14–16]. One limitation is potential discomfort when visible signal is emitted from sensors worn by a user. VLC is more suitable for downlinks when illumination and communication are coupled. In addition, the constraints on size and portability are not treated in these studies.

In this work, we propose an innovative scheme for ECG monitoring during physical activities that is:

- a portable device without cables on the chest and wirelessly transmitting
- based on optical wireless communications in infrared (IR) range

We present in Sect. 2 the optical wireless ECG monitoring system, its hardware description and the optical wireless frame format. Section 3 presents the methodology for evaluating the performance before discussing on the results. Section 4 concludes and discuss about future work.

2 Optical Wireless ECG Monitoring System

2.1 Hardware Realization

In order to transmit ECG data, the practical transmission system is designed with off-the-shelf components. The goal is to be able to transmit data while the person is making exercise inside a room, so with mobility. For this, we propose a configuration with spatial diversity including several optical receivers placed in the ceiling. They are at the corners of a central panel, which can be a lighting device as shown in Fig. 1 (a). In addition, we have fixed the receivers orientations to 45° thanks to pieces shown in Fig. 1 (b) in order to ensure good coverage area.

For the emission, we have designed a system able to be worn on electrodes integrated in a shirt, with the emitting diode oriented perpendicularly to the body (see Fig. 1 (c)). As we can see in this picture, the designed device has reduced dimensions in order to be worn upon a cardiac belt (space between electrodes is 5 cm). Therefore, the ECG signal amplitude may differ from classical ECG monitor.

The system has been designed considering the particularities of the optical receivers, which are integrated photo-detection modules (TSOP 34338), composed of PIN photo-detector and preamplifier. They are able to detect data frames when the signal is modulated following a subcarrier of 38 kHz with a minimum of 6 cycles per burst. The receiver peak wavelength detection is around 940 nm, it has a Field Of View (FOV) of 45° and a typical minimal irradiance of 100 $\mu W/m^2$ depending on the ambient noise level, which allows detecting very low power signal. The physical surface of PIN photodiode is 34.5 mm².

The output signal after the photo-detection module is directly connected to a FTDI module so that a computer can detect frames and evaluate packet error rate (PER) and save raw data. The serial transmission baud rate is chosen among normalized serial baud rates, and satisfying the fact that we must have at least 6 periods of the modulated signal at 38 kHz inside one useful digit. The raw data rate is consequently chosen at 4.8 kbps, which is the maximal one considering the 38 kHz sub-carrier.

With these considerations, the emitter is designed with a microcontroller generating two output signals; one is a Pulse Width Modulation (PWM) signal at 38 kHz, the other one the serial output at 4.8 kbps containing useful data. The outputs are linked to an electronic circuit making logical operation and driving an infrared Light Emitting Diode (LED) source (TSAL 6400) having a peak wavelength around 940 nm and half-power angle of 25°.

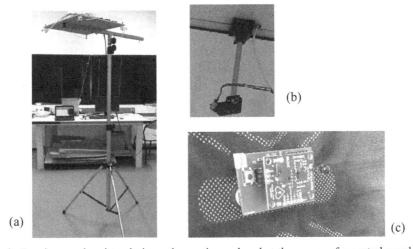

Fig. 1. Receivers and emitter designs: the receivers placed at the corner of a central panel (a), pieces for orientation of receivers (b), the emitter on the T-shirt (c)

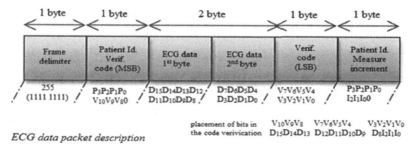

Fig. 2. Data packet description

2.2 Optical Wireless Frame Description

In order to be able to decode data at the receiver side and to count errors and lost packet, the frames have to follow a given format, with frame delimiters, packet numbering and redundant information. For this aim, we have developed a home-made protocol, defining a 6-byte frame as described in Fig. 2. Each sampled raw data of ECG are over 2 bytes, but we add a frame delimiter byte, a patient identifier code over 4 bits, a measure increment code from 0 to 7 (3 bits) for numbering the packets. Last transmitted data are a verification code calculated from raw data byte combination as well as redundant information of patient identifier. These two elements are used in order to detect erroneous data at reception side. We can notice that a more sophisticated protocol could be used if there was a constraint about data security or a multi-user context, but it is not treated in this study.

In order to ensure a low average IR transmitted power satisfying eye safety and considering the maximal forward current of the LED, we have made the choice to transmit data only 1/3 of time. For this, we add a delay between two transmitted frame equal to twice the time of transmission of one packet (here equal to 12.5 ms because of the baud rate of 4.8 kbps). Therefore, this means that the packet is transmitted each 37.5 ms. In addition, we have designed the electronic circuit to emit an instantaneous optical power of 51 mW. Taking into account the emitting time ratio of 33.3%, and the hypothesis of equiprobable data emission with the OOK modulation, the average IR transmitter power is around 4 mW.

3 Results and Performance

3.1 Performance Evaluation Principle

The computer serially receives data, and a Python program decides from redundant information whether there is an error of transmission or if at least one of the receivers correctly received the packet. Number of errors can be counted thanks to the packet number compared to the previous correctly decode packet number, but if there are more than 8 consecutive errors, it is not possible to correctly evaluate it. For this aim the python program also take into account time of arrival of correctly decode packets. More precisely, it computes the time that has passed between the previous correct packet and the correct one just received and compare it with the normal transmission time given by the equation:

$$T_T = \frac{Data(bits)}{Baudrate} + T_{delay} \tag{1}$$

T_{delay} is the delay after sending packet information (here twice the time of transmission, i.e. 25 ms). *Data* corresponds to the number of bits per packet (here 60 bits because of the 6 bytes serially transmitted, where each bytes is composed of 8 data bits, 1 start bit and 1 stop bit) and *Baudrate* is the serial data rate (here equal to 4.8 kbps). So in this configuration $T_T = 37.5$ ms.

In this case, at each received packet number i, the following quantity is determined:

$$nb^i_{lost} = \frac{\left(T^{(i)}_{pr} - T^{(i-1)}_{pr} - T_T\right)}{T_T} \tag{2}$$

$T^{(i)}_{pr}$ is the time of arrival of correctly received packet and $T^{(i-1)}_{pr}$ the time of arrival of the previous one. If nb^i_{lost} is not null, then it corresponds to the number of errors at this time of arrival. Finally, the PER for a transmission with n received packets and nb_{lost} packets is expressed as:

$$PER = \frac{nb_{lost}}{n + nb_{lost}} \tag{3}$$

where $nb_{lost} = \sum_{i=1}^{n} nb^i_{lost}$ is determined using (2).

The program computes this metric and prints this value and the number of errors at each received packet so that we can make experiments until a given number of errors is reached.

3.2 Results

The experiments consisted in walking in a room of size 6.6 m × 6.7 m wearing the optical wireless ECG device, with optical receivers in the center of the room at the corners of a central panel at a height of 2.5 m.

After 5 min, the PER converged to a value of 8.6%. This value is interesting because it is still possible to get a correct shaping of ECG as shown in Fig. 3. Note that Fig. 3 represents the amplitude of the ECG signal, normalized between 0 and 1023 as a function of the number of received sample, depending on the baud rate. For example, with the baud rate of 4.8 kbps, the sampling frequency is equal to 26 Hz, so the time between two samples is 38 ms. This value is sufficient to extract cardiac frequency but is too small to extract medical information from the ECG.

Moreover, we verified that in this case, with a packet loss of 8.6%, the heart rate is well evaluated. This has been validated by simultaneously wearing a reference heart rate monitor, i.e. Polar V800 cardiac belt evaluating R-R period and a connected watch.

Finally, in order to increase the sampling frequency, we have investigated same scheme using another receiver (TSOP 7000), working with a subcarrier of 455 kHz

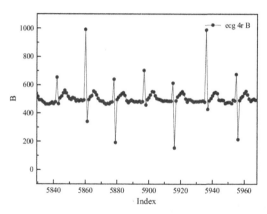

Fig. 3. Example of ECG from received data

instead of 38 kHz. In this case, it is possible to use baud rates from 4.8 kbps to 38.4 kbps. Increasing the baud rate allows reducing the packet emission time.

By keeping a delay equal to twice the packet emission time, the corresponding time between each packet T_T is determined in Table 1 in addition to the corresponding sampling frequency. We can note in this case that only the baud rate of 38.4 kbps permits reaching approximatively the minimal sampling frequency of 250 Hz.

The photodiode included in the TSOP 7000 receiver is not centered in the same wavelength as the previous one, but the sensitivity at 940 nm is 0.8. In addition, one limitation of using this device is the minimal irradiance able to be detected, which is 10 times higher. Therefore, experiments have been conducted with mobility in a smaller area inside the room, with radius 1.8 m from the center of the panel.

In order to transmit data with a subcarrier of 455 kHz, the transmitting device has been adapted, especially the LED (SFH 4346), logical gate and mosfet having lower response time have been used. In addition, a higher speed microcontroller has been used in order to be able to generate a precise PWM clock at 455 kHz.

The PER obtained for the different baud rate are shown in Table 2. We can note that the PER is higher when the baud rate increase, which can be a problem for an accurate ECG since a PER lower than 1% is recommended [2]. However, regarding the heart rate, we have verified that its evaluation is not impacted; even with 30% of lost packets, we are able to recover the correct heart rate value for these sampling frequencies.

Table 1. Sampling frequency corresponding to different baud rates at 455 kHz

Baudrate	T_T [ms]	Sampling frequency [Hz]
9600	18.7	53.3
19200	9.37	106.7
38400	4.7	213.3

Table 2. Experimental PER obtained with 455 kHz carrier frequency

Baudrate	PER, [%]
9600	12.6
19200	29.5
38400	26.5

4 Conclusions

The work presented in this paper answers to a problematic of distant monitoring and proposes a portable device able to perform distant ECG monitoring during physical activities. The proposed device is based on optical wireless communications and is designed using on-the-shelf components. More particularly the optical receiver working with a carrier frequency imposes constraints on raw data rate. Therefore, the sampling frequency in the proposed system is not sufficient for medical diagnosis of ECG but permits evaluating heart rate. Actually, we experimentally evaluate transmission quality when a person wearing the realized device is moving inside a room. We determine packet error rate of the ECG optical wireless transmission and we show that the heart rate is correctly evaluated even if packet error rate increases with data rate. Finally, our study shows the potentiality of using optical wireless for transmission and distant analysis of ECG data provided the development of specific component platforms.

Future work will consist in the integration of the device for being able to work at higher frequency with the 455 kHz subcarrier in order to reach the sampling frequency required for ECG evaluation. In addition, the choice of the LED should be optimized according the experimental results in order to lower the PER. Once the single device communication will be optimized, the next step will be to add other derivations for the ECG measurement, by placing electrodes at different places on the body. The aim will be to develop a system with independent active electrodes, each sending separately the measured value. In this case, a specific work will be done in order to deal with multiple access since the different electrodes send data simultaneously.

References

1. Kligfield, P., et al.: Recommendations for the standardization and interpretation of the electrocardiogram: Part I: the electrocardiogram and its technology: a scientific statement from the American Heart Association Electrocardiography and Arrhythmias Committee, Council on Clinical Cardiology; the American College of Cardiology Foundation; and the Heart Rhythm Society: endorsed by the International Society for Computerized Electrocardiology. Circulation 115(10): 1306–1324 13 March 2007. https://doi.org/10.1161/circulationaha.106. 180200. PMID 17322457
2. Dilmaghani, R.S., Bobarshad, H., Ghavami, M., Choobkar, S., Wolfe, C.: Wireless sensor networks for monitoring physiological signals of multiple patients. IEEE Trans. Biomed. Circu. Syst. 5(4), 347–356 (2011)
3. Alfarhan, K., Mashor, M., Rahman Mohd Saad, A.: A review of wireless ECG monitoring system design International Academy of Engineering and Medical Research (IAEMR) (2016)

4. Preejith, S.P., Dhinesh, R., Joseph, J., Sivaprakasam, M.: Wearable ECG platform for continuous cardiac monitoring. In: 2016 38th Annual International Conference of the IEEE Engineering in Medicine and Biology Society (EMBC), Orlando, FL, 2016, pp. 623–626 (2016)

5. Park, C., Chou, P.H., Bai, Y.,Matthews, R., Hibbs, A.: An ultra-wearable, wireless, low power ECG monitoring system. In: 2006 IEEE Biomedical Circuits and Systems Conference, London, pp. 241–244 (2006)

6. Chen, F., Wu, H., Hsu, P.L., Stronger, B., Sheridan, R., Ma, H.: SmartPad: a wireless, adhesive-electrode-free, autonomous ECG acquisition system. In: 2008 30th Annual International Conference of the IEEE Engineering in Medicine and Biology Society, vol. 2008, pp. 2345–8 (2008). http://doi.org/10.1109/IEMBS.2008.4649669

7. Al-Ahmadi, S., et al.: Multi-user visible light communications: state-of-the-art and future directions. IEEE Access **6**, 70555–70571 (2018)

8. Cogalan, T., Haas, H.: Why would 5G need optical wireless communications? In: 2017 IEEE 28th Annual International Symposium on Personal, Indoor, and Mobile Radio Communications (PIMRC), Montreal, QC, 2017, pp. 1–6 (2017)

9. Arnon, S.: Visible Light Communication, 1st edn. Cambridge University Press, New York (2015)

10. Dimitrov, S., Haas, H.: Principles of LED Light Communications Towards Networked Li-Fi, 1st edn. Cambridge University Press, UK (2015)

11. Ghassemlooy, Z., Alves, L.N., Zvanovec, S., Khalighi, M.-A.: Visible Light Communications: Theory and Applications, 1st edn. CRC Press, Inc., Boca Raton (2017)

12. Chevalier, L., Sahuguede, S., Julien-Vergonjanne, A.: Optical wireless links as an alternative to radio-frequency for medical body area networks. IEEE J. Sel. Areas Commun. **33**(9), 2002–2010 (2015). https://doi.org/10.1109/JSAC.2015.2432527

13. Behlouli, A., Combeau, P., Aveneau, L., Sahuguede, S., Julien-Vergonjanne, A.: Efficient simulation of optical wireless channel application to WBANs with MISO link. Procedia Comput. Sci. **40**, 190–197 (2014). ISSN 1877-0509, https://doi.org/10.1016/j.procs.2014.12.027

14. Tan, Y.Y. Jung,, S.-J., Chung, W.-Y.: Real time biomedical signal transmission of mixed ECG signal and patient information using visible light communication', in 2013 35th Annual International Conference of the IEEE Engineering in Medicine and Biology Society (EMBC), Osaka, 2013, pp. 4791–4794 (2013)

15. Yee-Yong, T., Wan-Young, C.: Mobile health monitoring system through visible light communication. Bio-Med. Mater. Eng. **6**, 3529–3538 (2014)

16. Al-Qahtani, A., et al.: A non-invasive remote health monitoring system using visible light communication. In: 2015 2nd International Symposium on Future Information and Communication Technologies for Ubiquitous HealthCare (Ubi-HealthTech), Beijing, China, 2015, pp. 1–3 (2015)

Emergency Patient's Arrivals Management Based on IoT and Discrete Simulation Using ARENA

Kaouter Karboub[1,2](✉), Tabaa Mohamed[3](✉), Fouad Moutaouakkil[2](✉),
Dellagi Sofiene[4](✉), and Abbas Dandache[4](✉)

[1] Research Foundation for Development and Innovation in Science (FRDISI),
HASSAN II University, Casablanca, Morocco
Kaouter.karboub@gmail.com
[2] Engineering National and High School of Electricity and Mechanic (ENSEM),
HASSAN II University, Casablanca, Morocco
fmoutaouakkil@hotmail.com
[3] Pluridisciplinary Laboratories of Research and Innovation (LPRI), EMSI Casablanca,
Casablanca, Morocco
med.tabaa@gmail.com
[4] Laboratory of Genie Industrial and Production of Metz (LGIPM),
Lorraine University, Metz, France
{sofiene.dellagi,abbas.dandache}@univ-lorraine.fr

Abstract. The healthcare ecosystem is now in a state of flux. Social, economic pressures beside demographic changes disrupt the balance of the health facilities. According to the Organization for Cooperation and Economic Development (OECD, 2004), "the last thirty years have been a period of change and of expansion for health systems". Currently, the major problem remains about controlling the ever-increasing health expenditures. Thus, Hospitals are faced with a triple constraint: cost, time and quality.

The current ecosystem must simultaneously integrate these constraints in order to offer the best possible service to patient in a minimum time and at the optimal patient's situation. We focus our interest on patients suffering from heart diseases, but still the global approach of the proposed model valid for other contexts. The proposed model is based on an extreme danger situation consisting of heart attacks.

In this paper, we aim to establish an embedded connectivity between heart diseases patients and their physicians. Real time monitoring plays an important role to establish that kind of connectivity.

In fact, the proposed simulation model is a Dynamic Stochastic Discrete model realized using ARENA software. All the results presented here are based on random data and can be replaced by real time data extracted and preprocessed using IoT sensors recording, and are referenced to a bi-objective function we are going to present in the following sections.

Keywords: Healthcare ecosystem · Time · Quality · Heart diseases · Embedded connectivity · Simulation · Emergency department · Dynamic Stochastic Discrete Model · Bi-objective function · IoT sensors

© Springer Nature Switzerland AG 2020
O. Habachi et al. (Eds.): UNet 2019, LNCS 12293, pp. 234–244, 2020.
https://doi.org/10.1007/978-3-030-58008-7_19

1 Introduction

The intensive population growth has considerably affect citizen's living level, and has remarkably limited their access to public services such as Health care. In 2016 and according to the World Health Organization, heart diseases was classified as the most significant death cause with 15.2 million deaths [1]. Beside the social impacts, this growth means a big increase of population problems including flows, transportation, equal education and treasure distribution. This high level of urbanization and population distribution is concentrated in the most developed cities. This is determined by several factors, the most significant are: new transportation modes, high speed rails, access to technology, education, and healthcare. Both citizens and governments need a continuous availability of public health care systems. To guarantee satisfaction and low mortality rates, those systems need to satisfy some efficiency and quality levels. To do so, health-care entities had developed different strategies that adopt strategic thinking, planning momentum and optimized dispatching rules for the current dynamic and coming health care systems. Effectively, the challenges related to managing incoming patients to the emergency department encourage adopting patients monitoring. The monitoring can be accomplished with many medical strip, wearable, implantable, invasive/non-invasive and ingestible sensors [2]. A variety of sensors can be used to measure many medical conditions like: heart rate, blood pressure, temperature and so on, in many stages of the rescue process, namely the ECG sensors. Additionally, many other constraints must be taken into consideration like staff availability, the material needed, allocation of beds, and waiting times. In fact, new technologies play a main rule to facilitate communication between all the healthcare ecosystem parts, mainly the Internet of Things (IoT). The IoT have shown their capacity to enhance the healthcare services quality. Based on a literature review, [3] has summarized the IoT applications in the healthcare field and oriented the next researches. Indeed many based IoT solutions has been proposed. For example: the smart rehabilitation introduced in [4] and [5]; this solution connects all the available resources over the internet, with patient, performing different activities like diagnosing and monitoring. In addition to many architectures used to analyze data [6]. Others focused on location using many wireless technologies like Bluetooth, RFID, Wi-Fi, ZigBee. One of these systems are Real Time Location Systems. They are used to track things. The Geographic Positioning System GPS is the most well-known of these examples [7]. Many other sensors have been developed to insure a continuous data acquisition about the patient situation. Although, ignoring the context aware factor is one of the important parameters that may lead to failure and resistance [8]. The main objective of Emergency Department is to provide immediate and accurate service once a demand is received. This requires a detailed study taking into account many parameters [9]. In this paper, Marco Amorim et al. proposed an intelligent dispatching algorithm that take into consideration the survival function which in case of emergencies, is time within what the patient can be saved.

The current article represents an ARENA model used to describe the problem and track flows. The paper contains three main sections; introduction, state of the art and the proposed model respectively, then the paper is ended by future perspectives and conclusion.

2 State of the Art

The technological development have generated many subfields like internet of things used today in many domains including the healthcare.

The health care system is a complicated ecosystem. The coordination between its different factors needs a comprehensive analyze to determine the link between them [10], effectively, in [11], the authors have divided the health care ecosystem into four layers. Micro including focal patients, clinicians, nursing staff and health professionals. Meso involving hospitals, clinics, local health support agencies. Macro containing state health authorities, professional associations, unions in health care sector and health insurers, and Mega level including the government agencies, health funding bodies, regulatory bodies and media. That ecosystem interacts and provides population evaluated services [12], and still the main objective of ED: Emergency Department is to provide an immediate and accurate medical service. However, achieving that objective is a big challenge all over the world due to the increasing demand for emergency consultations and the lack or poor access to primary care. This impacts negatively the quality of the service, and patient's satisfaction [13, 14].

Thus, the unpredictable number of the arrivals in ED is a big obstacle to manage staff and resources inside the hospital. Many researchers have worked on establishing forecasting strategies to predict the arrivals of patients, [15–20]. Ozgur M. Araz et al. [21], presents a predictive models (logistic regression, artificial neural network, decision tree, random forest, support vector machine and extreme gradient boosting (XGBoost)) for patient admission. Those models based on multiple years triage data collected once a patient arrives to emergency department. This model has for main aims to manage crowding and reduce the length of stay. To evaluate these models the researchers used many criteria like: AUC, model accuracy, positive predictive value and Gini statistics. As a result they determined the main variables leading to patient admission which are the case acuity, the arriving mode and age. Others used ARIMA models (Auto Regressive-Integrated Moving Average) [22], once patient's admitted, they can be moved to other departments.

Many studies focused on alternative bed management strategies impact. The authors in [23], developed a multi-objectives optimization model to determine the parameters involved in decision making. Despite the presence of many conflicting objectives, the study took into consideration both sides; the operational and tactical. In addition to the impact of the supply chain inside the hospital that includes about 40% of the hospital budget [24].

Supervision systems like Mobile Emergency System (MES) presented in [25] play an important role to monitor ED admitted patients. Those systems can help alerting physicians and compromise clinical staff limitations. This solution establishes a continued communication between inpatients and health care providers. In fact IoT emerging technologies give hope for patients. In both, homecare and in hospitals, Body Sensor Networks (BSN) is one of the potential applications of WSN [26]. Presents a model tested by neural network (NN), using IoT, cloud computing and data mining. In order to provide an instant information of the heart state using electrocardiogram (ECG). In the model proposed in [27], the data generated from the ECG is directly transmitted to the IoT cloud using WIFI. The data collected from ECG sensors is stored in the IoT cloud.

To have instant access to the needed data, Hypertext Transfer Protocol (HTTP) and Message Queuing Telemetry Transport (MQTP) beside Graphical User Interface were used. In [28], the authors have presented another model called UbiMon Body Sensor Network Architecture. Containing six interconnecting parts, starting with sensors, remote sensing units, local processing units, the central server, the patient database, and the workstations. The UbiMon is a project developed to monitor the patient situation continuously, using many sensors and devices to provide several medical indicators as the heart rate and the blood oxygen saturation, [29]. The UbiMon have shown its efficiency to help taking decisions in high patient's risk levels [30].

3 The Discrete Events Simulation Proposed Model

3.1 The Mathematical Formulation

The health care ecosystem consists of physicians, nurses, hospital staff, ambulances, suppliers and most importantly patients.

To formulate the problem, we consider the following variables:

- Pi: the variable equal to 1 is at least one patient arrives at the hospital in time t.
- Sl: the length of the hospital stuff.
- Wi: the time of the patient i before being diagnosed by a physician
- μi: the variable equal to 1 if patient i had been received at least once for the same pathology, heart diseases in this case.
- βki: equal to 1 if the patient i is assessed to a physician j associated with department k.
- Aσ(t + Dt), i: equal to 1 if patient i is assessed to a bed with σ(t + Dti) specifications, expressed as a function of the department, pathology gender and age, and Dti the time between a patient being diagnosed and assessed to a bed.
- LoSi: the Length of Stay of patient i

The length of the patient's residence can be divided in our approach into three main parts; the waiting time, the time needed for diagnostic and the length of stay.

Using those variables, we express the bi objective function as shown below:

$$F : \begin{cases} Min \sum_{I+1}^{n} Pi(t) * \mu i * wi \\ \sum_{i=1}^{n} \sum_{j \in Sl} \beta jki * A\sigma(t)i * Pi \end{cases} \tag{1}$$

Mathematically expressed above, the objective of our project is to minimize the waiting time of a patient arriving at time t, and to maximize the number of patients benefiting from the physicians diagnostics and the assessment to adequate bed.

We introduce a **Dynamic Stochastic Discrete** simulation model using ARENA software.

The Proposed Simulation Model

We propose a **Dynamic Stochastic Discrete Model** using ARENA simulation software. The model is an illustration of patient's traditional workflow meant to be used as optimization target.

To imitate the uncertainty and the variability of the patient's arrivals, we based our model on random data. The healthcare structure contains 30 physicians and an infinity of arriving patients. We are going to explain the model's details further. The table below presents the distributions we generated using ARENA Input Analyser (Table 1).

Table 1. The entities distributions

Entity	Distribution characteristics
Entity 1 (Patients arrivals)	Distribution: Weibull Expression: $-0.5 + $ WEIB(6.56, 1.91) Quadratic error: 0.004486 Statistic test: 11.7 Number of data points: 168 Min value: 0 Maximum value: 19 Mean value: 5.32 Std Dev: 3.17
Entity 2 (doctors availability)	Distribution: Weibull Expression: $-0.5 + $ WEIB(9.43, 1.47) Quadratic error: 0.006018 Statistic test: 10.9 Number of data points: 175 Min value: 0 Maximum value: 28 Mean value: 8.87 Std Dev: 5.87
Entity 3 (ambulances availability)	Distribution: Beta Expression: $-0.5 + 10 * $ BETA(1.33, 1.21) Quadratic error: 0.007043 Statistic test: 10.9 Number of data points: 168 Min value: 0 Maximum value: 9 Mean value: 4.73 Std Dev: 2.65

Using the distributions above, we built the following simulation model represented in Fig. 1:

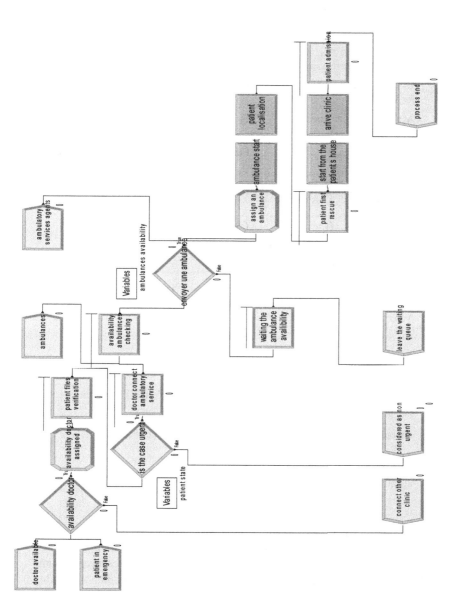

Fig. 1. The proposed simulation model

We resume in the following table, the most important parameters defined in The ARENA model description (Table 2).

Table 2. Entities/queues/resources/variables used in the proposed simulation model

Queues	*Name*		*Type*	*Allocation*
	Patient files verification		First in First Out	Wait
	Doctor connects ambulatory service		First in First Out	Transfer
	Availability ambulances checking		First in First Out	Wait
	Patient first rescue		First in First Out	Added value
	Patient admission		First in First Out	Added value
	Waiting the ambulance availability		First in First Out	Wait
Resources	*Name*	*Type*	*Capacity*	*Per use*
	Doctor	Fixed capacity	30	80
	Clinical stuff	Fixed capacity	100	80
	Ambulatory service agent	Fixed capacity	30	80
Variables	*Name*		*Data type*	
	Patient state		Real	
	Ambulances availability		String	
Entities	*Name*		*Expression*	*Max arrivals*
	Patient in emergency		$-0.5 + \text{WEIB}(6.56, 1.91)$	Infinite
	Doctor available		$-0.5 + \text{WEIB}(9.43, 1.47)$	30
	Ambulances		$-0.5 + 10 * \text{BETA}(1.33, 1.21)$	10
	Ambulatory services agents		$\text{UNIF}(1.6, 3.3)$	30

After compilation we get the following results, taking into account the assumption that the time taken by each resource to accomplish the required operation is standard and does not change while doing the next operation, and that all resources have the same attitude and the same allure. The following figure summarizes the results concerning the entities taken into consideration, namely: patients, doctors, nurses and ambulances services agents checked in every hour during 12 h (Fig. 2):

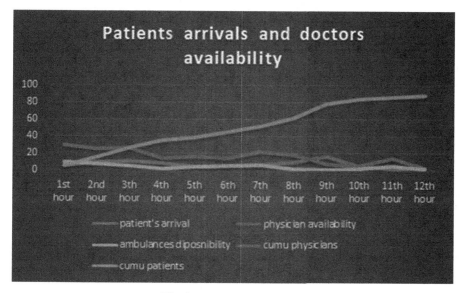

Fig. 2. The simulation results

The previous results, make it clear that we should focus on two main point:

- The synchronization of the operations done by the staff in the hospital and outside the hospital, this can be done by establishing a standard of the duration of the elementary operation, and this can also take into account the distribution of the adequate resources, as a result we can minimize the waiting time and make the model smoother.
- It made clear the necessity to predict the patient's arrival.

3.2 Discussion

Modeling and simulation had been a great way to support decision making across time. Many researches have studied many models using many approaches to predict, analyze or to get over the missing data [31, 32]. Queuing theory (QT) is a classic methodology based on mathematical models, developed to obtain the closest shape parameters that allow researches, engineers to design the needed performance metrics. Those models has been widely used in the patient flow management [33]. However, using the queuing models directly in the health care seems to be inappropriate due to the complexity and high level of needed accuracy requested in this field [34]. The assumptions used in QT can be useful to estimate the waiting time and the queue length. Although, they still not close enough to the operational process, a thing that can be done by discrete event simulation. In [35], authors developed a forecasting model for arrival patients in the emergency department. They consider an arrival patient once he/she go through the administrative registration process neglecting if he/she going to leave or not and limiting the length of stay in 24 h. The patient may leave the process for two reasons. First, the patient is treated and sent back home. Second, he is reoriented to another healthcare entity.

In the proposed model (Fig. 1), we consider an arrival when a medical resource is taken, it can be doctor, nurse, ambulance, and administrative staff. This new definition allows us to make it evident the correlation between arrivals patients and resources needed to insure a high quality service level. The patient is equipped with sensors enabling a continuous communication between the health care provider and the patient using his User Interface, so he can trigger a rescue demand. The availability of the doctor defines if a patient is diagnosed and admitted into the ED. If doctor available, he examines the patient files and judge if the case is urgent or not. If doctor not available the patient is reoriented to another health care entity. In case the situation is critical, the doctor send a call for ambulatory vehicle that is sent directly to the patient location. If ambulance is available the patient is served, transported to hospital, in the hidden time of transportation, the appropriate equipment and resources judged by the doctor are prepared to receive the patient. In this way we eliminate a big source of waste which is the triage phase. We should also notice that the registration is the last phase of the rescue process, this choice is done for two reason: the patient files already exist as we treating pathologic patients in emergency cases, second to save time leaving the administrative registration in the end and to keep the last updated version of the patient's files.

4   Perspectives and Conclusion

In the future works, we aim to develop an AI model to predict the patient's arrival based on this real time ECG recordings.

In the current paper, we presented in a Dynamic Stochastic simulation model to track the patient trajectory and analyze the flows inside and outside the hospital. This model is a referential model, further achievements will be tasted on this model to compare the gain in every approach. The global solution will be used to determine the couple of decisions variable in (1) that optimize the maximum our objective which is to minimize the mortality rate.

To predict the patient's arrivals, we should understand and be aware of the context where those patients live (environmental factors, access to medical services, behavior, nutrition habits …) this can help us provide an estimation of the arriving patient to emergency department and build an intelligent model capable to support the decision making. Thus, the prediction should not concentrate on only arrivals patients as all healthcare ecosystem parts can impact the results of this prediction.

References

1. The top 10 causes of death, the World Health Organization. Accessed 02 June 2019
2. MarketsandMarkets homepage, Medical Sensors Market by Sensor Type (2017)
3. Yin, Y., Zeng, Y., Chen, X., Fan, Y.: The internet of things in healthcare: an overview. J. Ind. Inf. Integr. **1**, 3–13 (2016)
4. Rohokale, V.M., Prasad, N.R., Prasad, R.A.: cooperative Internet of Things (IoT) for rural healthcare monitoring and control. In: Proceedings of the 2nd International Conference on wireless Communication, Vehicular Technology, Information Theory and Aerospace & Electronics Systems Technology (Wireless VITAE), pp. 1–6. IEEE, February 2011

5. Bertholdo, L.M., Granville, L.Z., Arbiza, L.M.R., Carbone, F., Marotta, M., de Santanna, J.J.C: Internet of Things in healthcare: Interoperability and security issues. In: Proceedings of the IEEE International Conference on Communications (ICC), pp. 6121–6125. IEEE, June 2012

6. Plageras, A., Psannis, K., Stergiou, C., Wang, H., Gupta, B.: Efficient IoT-based sensor BIG Data collection–processing and analysis in smart buildings. Fut. Gener. Comput. Syst. **82**, 349–357 (2018)

7. Peng, R., Sichitiu, M.L.: Angle of arrival localization for wireless sensor networks. In: Proceedings of the 3rd Annual IEEE Communications Society on Sensor and Ad Hoc Communications and Networks, SECON 2006, vol. 1, pp. 374–382. IEEE, September 2006

8. Neto, A., Junior, J., Neuman, J., Cerqueira, E.: Context-aware eHealth information approach for the Brazilian primary healthcare system. In: 2013 IEEE 15th International Conference on eHealth Networking, Applications and Services (Healthcom 2013), pp. 274–276. IEEE (2013)

9. Amorim, M., Ferreira, S., Couto, A.: Emergency medical service response: analyzing vehicle dispatching rules. Transp. Res. Rec. J. Transp. Res. Board **2672**(32), 10–21 (2018)

10. Chaerul, M., Tanaka, M., Shekdar, A.V.: A system dynamics approach for hospital waste management. Waste Manage. **28**(2), 442–449 (2008)

11. Frow, P., McColl-Kennedy, J.R., Payne, A.: Co-creation practices: their role in shaping a health care ecosystem. Ind. Market. Manage. **56**, 24–39 (2016)

12. Bernstein, S.L., et al.: The effect of emergency department crowding on clinically oriented outcomes, Society for Academic Emergency Medicine, Emergency Department Crowding Task Force, Acad Emerg Med (2009)

13. Boyle, J., Jessup, M., Crilly, J., Green, D., Lind, J., Wallis, M., Fitzgerald, G.: Predicting emergency department admissions. Emergency Med. J. **29**(5), 358–365 (2010)

14. Hoot, N.R., et al.: Forecasting emergency department crowding: an external. Multicenter Eval. Ann. Emerg. Med. **54**(4), 514–522 (2009)

15. Billings, J., Georghiou, T., Blunt, I., Bardsley, M.: Choosing a model to predict hospital admission: an observational study of new variants of predictive models for case finding. BMJ Open **3**(8), e003352 (2013)

16. Diaz, J., et al.: A model for forecasting emergency hospital admissions: effect of environmental variables. J. Environ. Health **64**(3), 9–15 (2001)

17. Hoot, N.R., Zhou, C., Jones, I., Aronsky, D.: Measuring and forecasting emergency department crowding in real time. Ann. Emerg. Med. **49**(6), 747–755 (2007)

18. Hoot, N.R., Aronsky, D.: Systematic review of emergency department crowding: causes. Effects Solut. Ann. Emerg. Med. **52**(2), 126–136 (2008)

19. Jones, S.A., Joy, M.P.: Forecasting demand of emergency care. Health Care Manage. Sci. **5**(4), 297–305 (2002)

20. Kam, H.J., Sung, J.O., Park, R.W.: Prediction of daily patient numbers for a regional emergency medical center using time series analysis. Healthcare Informatics Research **16**(3), 158–165 (2010)

21. Araz, O.M., Olson, D., Ramirez-Nafarrate, A.: Predictive analytics for hospital admissions from the emergency department using triage information (2018)

22. Carvalho-Silva, M., Monteiro, M.T.T., de Sa-Soares, F., Doria-N'obrega, S.: Assessment of forecasting models for patient's arrival at Emergency Department. Oper. Res.r Health Care **18**, 112–118 (2018)

23. Landa, P., Sonnessa, M., Tanfani, E., Testi, A.: Multiobjective bed management considering emergency and elective patient flows. Int. Trans. Oper. Res. 25(1), 91–110 (2018)

24. Hospitals & Health networks Strategic supply chain management. ProQuest Central (2011)

25. Chiou, S.-Y., Liao, Z.: A real-time, automated and privacy preserving mobile emergency-medical-service network for informing the closest rescuer to rapidly support mobile-emergency-call victims (2018)
26. Nachman, L.: "Imote2" TinyOS Technology Exchange II (2005)
27. Yahyaie, M., Tarokh, M.J., Mahmoodyar, M.A.: Use of Internet of Things to Provide a New Model for Remote Heart Attack Prediction (2018)
28. Van Laerhoven, K., et al.: Medical Healthcare Monitoring with Wearable and Implantable Sensors (2004)
29. Kumar, A., Kumar, A.: Evolution of cloud to fog computing | RankWatch Blog, Rankwatch.com, 2016. Accessed 19 May 2019
30. Al-Janabi, S., AlShourbaji, I., Shojafar, M., Shamshirband, S.: Survey of main challenges (security and privacy) in wireless body area networks for healthcare applications Egyptian In- form. J. (2016)
31. Hu, X., Barnes, S., Golden, B.: Applying queueing theory to the study of emergency department operations: a survey and a discussion of comparable simulation studies. Int. Trans. Oper. Res. (2017)
32. Rosenblatt, A., Gravenor, S., Gurvich, I., Van Mieghem, J., Kannan Mutharasan, R.: Optimizing emergency room throughput for cardiac telemetry patients: a queuing theory approach (2018)
33. Palvannan, R.K., Teow, K.L.: Queueing for healthcare. J. Med. Syst. **36**, 541–547 (2012)
34. Haghighinejad, H., Kharazmi, E., Hatam, N, Yousefi, S, Hesami, S., et al.: Using queuing theory and simulation modelling to reduce waiting times in an iranian emergency department. Int. J. Community Based Nurs. Midwifery **4**(1), 11–26 (2016)
35. Kim, K., Lee, C., O'Leary, K.J., Rosenauer, S., Mehrotra, S.: Predicting patient volumes in hospital medicine: a comparative study of different time series forecasting methods. 23 January 2014

MSND: Multicast Software Defined Network Based Solution to Multicast Tree Construction

Youssef Baddi[1]([✉]), Anass Sebbar[2,3], Karim Zkik[2], Mohammed Boulmalf[2], and Mohamed Dafir Ech-Cherif El Kettani[3]

[1] STIC, ESTSB-Chouaib Doukkali University, El Jadida, Morocco
baddi.y@ucd.ac.ma
[2] Université Internationale de Rabat, FIL, TICLab, Rabat, Morocco
[3] ENSIAS—University Mohammed V Rabat, Rabat, Morocco

Abstract. Internet services provider propose increasingly numerous applications depend on multicast communication patterns, such as database synchronization, newsletter updates, video, and audio group streaming. In spite of, IP multicast is frequently suffered from scaling and instability, which causes internet provider administrators to avoid its utilization. While different multicast protocols in all layers, such as routing and switching, have been created by researchers and providers, through several years, to focus on security, congestion control, and scalability, our work aims to scale multicast regarding the number of supported multicast groups and sources in networks and multicast session. In this paper, we propose an SDN-based IP multicast tree construction module to support and deploy multicast communication-based services. To do so, we take advantage to compute multicast tree, in the centralized way, to achieve a dynamic tree construction algorithm that helps SDN controller to dynamically construct and adjust an optimal multicast tree in all multicast session. The experiments studies on the prototype and simulation system exhibit that our solution gives better performance contrasted with the traditional multicast tree construction algorithms, better also to SDN-based solution proposed in the literature, regarding multicast tree cost, delay, delay variation, and multicast tree construction delay.

Keywords: SDN · Multicast IP · MSDN · VNS-RP · Multicast routing

1 Introduction

Group communication technologies enable internet customers to profit from a wide range of internet services, we can note usually, newsletter distribution, database synchronization, and audio or video-conferences [1]. Considering group communication technologies, multicast IP routing offers effective methods to help these kinds of communication, since it to a great extent saves the data transmission bandwidth of the network and decreases heap of the distinctive servers that have these diverse internet services.

The use of IP multicast generally suffers from several issues such as scalability and stability [2–4] which hinders its deployment. Despite his apparent benefits, network operators generally avoid its use. The main problem of using IP multicast in traditional

© Springer Nature Switzerland AG 2020
O. Habachi et al. (Eds.): UNet 2019, LNCS 12293, pp. 245–256, 2020.
https://doi.org/10.1007/978-3-030-58008-7_20

network is that each node in the network (Servers and routers in the network) must support the topology discovery function and that each node must compute adjust multicast tree.

To get around these problems, several multicast protocols have been designed and implemented by researchers and developers of providers through several years to address security, congestion control, and scalability. Despite the different results presented by this research, the scalability problem still persists and the used protocols require a large number of multicast requests. These drawbacks emerge on the grounds that switches and routers support restricted and finite number of multicast routes in their forwarding tables. This problem is increased by the fact that multicast addresses cannot be assigned in a centralized manner and therefore cannot be aggregated hierarchically using traditional networks. To avoid this problem, we point in this work to scale multicast regarding multicast groups size so as to facilitates the use of multicast IP mode in data centers.

Software Defined Networks (SDNs) is a new centralized network architecture model that enables network nodes to be centrally managed. This new emergent network design provide several advantages in terms of strategy, adaptability, scalability, and agility, making it simple to support all kinds of network technologies, internet services and applications. The SDN architecture is characterized by the separation of data and control planes, by several specific network devices (Controllers, OpenFlow-Switches) and by several protocols (OpenFlow, ForCES, OVSDB, BGP) for data flow and data management.

SDN architecture is composed of two components: SDN Switches and SDN controller. SDN Switches are only responsible for data forwarding and the SDN controller are responsible for deploying flow management rules. This centralized architecture will then allow us to have an overview of the whole network and the calculation of multicast trees will be done once and centrally at the level of the control plane.

In this paper, we propose an SDN-based IP multicast tree construction to support and deploy multicast communication-based services. To do so, we design a dynamic multicast tree construction algorithm that helps SDN controller to dynamically calculate and adjust multicast tree. So, we are going to design a multicasting controller that support multicast tree construction module, a multicast groups management and a Tree reconstruction and update module.

The rest of the paper is organized as follows. In Sect. 2, we present some background and terminologies and we describe SDN architectures and IP Multicast protocols. In Sect. 3, we present some related works. In Sect. 4, we present our proposed model and then we made a performance study to proof the efficiency of our model. Finally, in Sect. 5, we conclude the paper.

2   Background and Terminology

2.1   Multicast IP

Traditional IP communication can be achieved in two ways: one is based on unicast communication; the other based on broadcasting communication. Traditional unicast and broadcast communication methods cannot effectively solve the group communication problem. Multicast IP refers to sending a packet to a certain set of nodes in the form of best-effort transmission in an IP network. The basic idea is that the source hosts such as multicast sources send only one data. The destination address is the address of the multicast group; all recipients in the multicast group can receive the same copy of data and only hosts in the multicast group can receive the data, but other hosts cannot receive the data. The multicast router in this case is responsible to duplicate the data as the number of receivers in the multicast group. This IP multicast is used for generally to stream media and some other network applications and services. Intra-domain multicast routing protocols are classified into two categories, based on the receiver's topology distribution, as shown in Fig. 1, as: dense mode or sparse mode.

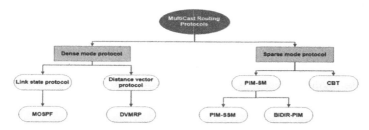

Fig. 1. Multicast routing protocols

Classical IP multicast has faced variety of challenges from its initiation, first by Deering [5]. Many limitations are known by researchers from totally different points of read in the literature [6, 7]. Though these limitations are known quite a very long time ago, most of them are still present and so relevant to check.

2.2 SDN

The term Software-Defined Networking (SDN) was firstly designed to extend ideas and work across OpenFlow protocol at Stanford University [8]. Software-Defined Networking (SDN) refers to network's where the state of the transfer in the data plane is changed by separating the remote-control plane from the first.

The figure shows the standard architecture of the SDN and the separation between the different planes:

- The data plane is made up of network devices responsible only for data transmission.
- The control plane manages the flow and consists of three fundamental abstractions: (1) transmission, (2) distribution and (3) specification.

The emergence of programmable SDN networks, use OpenFlow protocol in order to communicate data plan to control plan, for that many new data transfer features and routine protocol have become more flexible [9], such us multi-Pathing, flooding, link aggregation, fast rerouting and transfer of multiple stream to a next common hop.

The SDN architecture come as a big revolution in the field of computer networks, because they give many benefits and offers the opportunity to manage easily the infrastructure equipment layer, deployed applications and other functionalities. On the other hand, the achievement of SDN architecture is not clear and not deployed well, it requires great expertise; finally manage the network infrastructure, especially when implementing multicast techniques in an SDN environment.

2.3 Multicast SDN

Though the increase of multicasting appearances promising, with all its advantages, its use in total production is still limited due to many limitations that hinder large-scale deployment, some of this limitation are: the traditional architecture of the Internet, which is intended to provide best-effort forward for one-to-one communication, balancing minimum delay against other problem, reliability, scalability and quality of service.

In other side, the SDN promises a high level of flexibility and ease for network configuration by leveraging the concepts of control data plane separation and network programmability. This kind of conception provide a large range of possible optimizations and a flexible approach to multicast IP and multicast routing.

SDN technique provides new power in the multicast tree management, on behalf of using decentralized management protocols such us IGMP in IPv4 and MLD in IPv6, and multicast routing protocols to maintain multicast tree and forward multicast data. SDN offers new opportunities for global visibility and programmatic control in network switches.

3 State of Art

Using multicast techniques in SDN environment is a real challenge for network experts. Indeed, SDN's centralized architecture and intelligence lack at the data plan level (more precisely at the level of open flow switches) makes the exchange of information the information needed for the construction of multi-diffusion trees very difficult. In this sense, several works have proposed countermeasures and new protocols to bypass this issue. Marcondes et al. [10] proposed Castflow, an SDN-driven approach to IP multicast routing. The purpose of using Castflow, on the opposite to existing IP multicast protocols, is to limit the number of exchanges between different nodes and flooding messages when building the multicast tree taking advantages of the centralization of SDN architectures. Despite the encouraging results of the use of this framework but it is only suitable for small SDN networks (with a limited number of nodes). When implemented in a large network the results are not very conclusive.

In order to develop adopted solutions to larger SDN networks and to offer more scalability based on Dijkstra's Algorithm [7]. Jiang et al. Proposed an extended Dijkstra's Algorithm named Extended Dijkstra Short Path Algorithm (EDSPA) [11]. The basic principle of their works is to lighten the controller operations and increase network performance by using some load balancing technics. To do so, authors calculate the best route by using edges' weights like main parameters and by using virtual IP (VIP), when implementing their multicast tree construction algorithm EDSPA (Fig. 2).

When building multicast trees, it is very important to take into consideration the nature of the data and their different specificities (volumes, protocols and number of flow entries). It is clear that it is more difficult to deal with multiples flow entries connections. To bypass these issues Lin et al. [12] proposed a Locality-Aware Multicast Approach named (LAMA) which aims to reduce latency and calculation operations when constructing the multicast distribution tree. To do so, authors use their algorithm LAMA to reduce the number of flow entries and communications between SDN nodes and components when using video streaming packets.

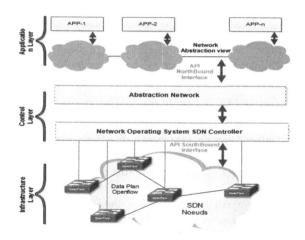

Fig. 2. SDN architecture

Considering scalability issues, Huang et al. [13] proposed BAERA (Branch Aware Edge Reduction Algorithm) an algorithm that aims to provide more scalability when using multicast technics. The main idea of their works is to reduce and optimize the number of branch and edge nodes when construction a multicast tree. To do so, authors use their algorithm to solve approximately the NP-hard problem Steiner tree ST.

The construction of multicast trees in the SDN is a very delicate operation. This is due to the fact that we must take into account the diversity of existing access paths in SDN architectures (especially from the control plane to the data plan). To address this issue, Iyer et al. [14] presented a multicast tree construction adapted to SDN networks. The authors propose the use of a new routing algorithm called Avra Routing Algorithm (AvRA) that aims to minimize the size of the routing tree in SDNs. Their approach is to look for the shortest path to the existing multicast tree at the level of all adjacent neighboring nodes of a multicast receiver that is trying to join the multicast tree.

The research work proposed in the literature offers several solutions that make it possible to use multicast techniques in SDN environments. However, these works do not deal with all the features of multicast routing protocols, namely: Group Management and Event Management membership such as group event management. In addition, the proposed solutions do not take into consideration the quality of service (delay, variation delay) and many of these algorithms are based on the algorithm dijekstra which is not effective when used in a centralized environment (from single node to multiple nodes) such as SDN architectures.

The purpose of our work is based on the following research questions:

RQ1: What is the best way to implement a multicast routing solution suitable for SDN architectures without compromising its operations.

RQ2: How to develop a generic and global solution that covers all the features of a multicast routing protocol (multicast management group, tree construction and event management group) without favoring a feature at the expense of others.

RQ3: How to develop a flexible and highly scalable solution that allows you to embed all other existing multicast techniques and algorithms.

4   Proposed Solution

4.1   Overview

Multicast-SDN send out message packet to a plurality of receivers in this case OpenFlow switches. The Fig. 3 show the network structure and transfer flow in the data plan, then transmitted the packet into SDN controller. In order to illustrate the different modules of multicast SDN.

In these subsections, we will describe in detail the implementations of the proposed solution as a series of modules, we start by tree construction module and group events Management module.

4.2   Multicast Tree Construction Module

The main function of this module is to ensure constructing Multicast Tree MT when the controller needs to start a new multicast session. In current work we use VNS algorithm to construct the multicast tree. The main motivation behind the use of the VNS search algorithm to solve Multicast Tree construction problem is the use of several neighborhoods structures to explore systematically different neighborhood structures from the set of sources to all destination set nodes of a specific multicast group.

Variable neighborhood search algorithm applied to multicast tree construction problem try to resolve three facts, firstly when creating the optimal multicast tree by combining all path from any source to all receivers, if this path P is a local minimum path for one neighborhood structure N is not necessarily a local minimum path with another one N'. if this path l is a global minimum path, it must be a local minimum path for all possible neighborhood structures. If not, links in local minimum paths for all neighborhood structures is relatively close and located in the same place.

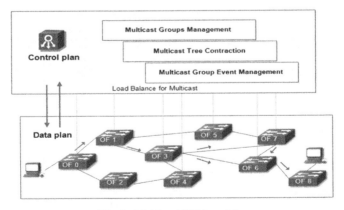

Fig. 3. Multicast SDN modules

In the next sections, we provide a detailed description of the Variable Neighborhood Search algorithm for Multicast Tree construction Problem with multicast tree cost, delay and delay variation guarantee used in MSDN solution, and his three sub-modules: the initialization process, Stopping conditions sub-module and the shaking sub-module.

4.3 VNS Algorithm

In these subsections, we will describe Variable Neighborhood Search (VNS) first proposed by Mladenović and Hansen [5, 15], instead of performing a local search method follow a trajectory, VNS is a meta-heuristics algorithm, which investigates increasingly multiple neighborhoods structures starting from a selected optimal local solution, and jumps from fined optimal local solution to a new one if and only if an improvement has been made.

The basic idea of VNS algorithm is a systematic change of arbitrarily chosen neighborhoods structures, which can vary in size, but usually with increasing cardinality, using in each iteration a local search algorithm [16], like Hill Climbing, Simulated annealing, tabu search algorithm …

In order to adapt the VNS algorithm to our proposed solution, a first step start by generating a set of neighborhoods structures and then creating an underlying initial solution, beginning from this initial possible optimal solution, a supposed shaking step is performed by randomly choosing another solution from the first next neighborhood structure. This is followed by applying a local search algorithm as an iterative enhancement algorithm. These steps are repeated so far as a new better optimal solution is found. Otherwise, we move to the next neighborhood structure and executes the shaking step followed by the iterative enhancement. In the event that another optimal solution is discovered we begin with the first neighborhood structure; else we continue with the next neighborhood structure, and so forth.

The basic VNS steps are first defined by Mladenović and Hansen in [5, 15], and adapted to our model, these steps are presented in Fig. 4, which presents the process of the algorithm as follows:

- Initialization: Create the set of neighborhood structures N_k, where k in interval [1, k_{max}], that will be used in each local search algorithm; select an initial solution S from the first neighborhood structure N_1; fix a stopping condition;
- Loop over the following instructions until the fixed stopping condition is met:

  (1) Set k $\leftarrow$ 1
  (2) Loop over the following steps until k = k_{max}

- Shaking phase: Randomly generate a point S' from the kth neighborhood of S (S' $\leftarrow$ $N_k[S]$)
- Local search: start a local search algorithm with this optimal solution S' as initial solution to find another better local optimum solution S'';

Fig. 4. VNS-based algorithm process

- stopping condition: in this phase the algorithm response to question: Move or not?. If this point S" is better than the first one (S), the algorithm move to this one (S ← S"), and continue the search with first neighborhood structure of this solution (k ← 1); otherwise, jump to the next neighborhood structure by setting k ← k + 1;

where N_k (k in [1, kmax]) is the finite set of pre-selected neighborhoods. The global stopping condition can be a limit on CPU time, a limit on the number of iterations, or a limit on the number of iterations where there is no optimal solution improvement.

4.4 Mathematical Modeling

A network topology mathematically can be formulated as a graph structure, G, this graph is generally directed and full connected. The graph structure G is composed by two elements N and L, N is a finite set of nodes and L is a set of links connecting the nodes. A link $l \in L$ connecting two neighboring nodes $n_1 \in N$ and $n_2 \in N$ is to be denoted by l (n_1, n_2). We associate to each link l two positive real values: a cost function C and a Delay function D. Where C (l) = C (l $[n_1, n_2]$) represents link cost, which may be any degree or quota of resource exploitation). The delay function D (l) = D (l $[n_1, n_2]$) is defined as the time needed by the multicast traffic to be forwarded between n_1 to n_2, including any multicast traffic handle like switching, queuing, transmission and reproduction delays. Using these two functions, we associate for each path P (n_0, n_n) = (l $[n_0, n_1]$, l $[n_1, n_2]$, ..., l $[n_{n-1}, n_n]$) in the network two metrics, we can formulate this two function as follows where F function can be C or D:

$$F(l(n_n, n_n)) = \sum_{i=0}^{n} F(l(n_i, n_{i+1})) \tag{1}$$

Multicast IP technology as it create a logical topology under network, the Multicast Tree M_T (S, C, D) can be modeled as a logical sub-graph of G covering all sources nodes S (S ⊂ N) and all destinations nodes D (D ⊂ N) with a selected core node C.

In Shared Tree ST-based protocols all sources node must forward in unicast mode the multicast data to selected core node C, after that, the core node transmit this multicast data to all receivers through the Shared Tree. To formulate the transmission of multicast data via these two parts divided by the core node C, we use the following equation, where F function can be both cost function and delay function:

$$F(M_T(S, C, D)) = \sum_{s \in S} F(l(s, C)) + \sum_{d \in D} F(l(C, d)) \tag{2}$$

Multicast IP-based application rely especially in the delay variation, we introduce a Delay Variation (7) function defined as the difference between the maximum (5) and minimum (6) end-to-end delays along the multicast tree from the source to all destinations nodes and is calculated as follows:

$$D_V = Max(D(M_T(S, C, D))) - Min(D(M_T(S, C, D))) \tag{3}$$

Multicast tree construction problem tries to find an optimal multicast tree with an optimal core node C in the network with an optimal function Opt_F by minimizing in the first time the cost function then a Delay and delay variation bound as follows:

$$Opt\_F(C, M_T) = \begin{cases} Min(C(M_T(S, C, D))) \\ D < \alpha \\ D_V < \beta \end{cases} \tag{4}$$

4.5 Initial Solution

In our adaptation of VNS algorithm to select the best first multicast path, we distinct two selection approaches to choose an optimal multicast path as an initial solution. The first one is the simplest, which randomly chose an initial path from all possible path in the network topology, the second one allows to reduce the area of selection using only the set of existing paths from the source and first multicast receiver explicitly join the multicast session.

4.6 Neighborhood Structures Sub-module

The strangeness source of the variable neighborhood search algorithm is because the fact that the global optimal solution found is the optimal solution of all neighborhood structures, wherever, in other proposition, an optimum with respect to a neighborhood is not necessarily optimal with respect to another [17].

Opposed to local search algorithms, which use by default a single neighborhood structure with one trajectory search, VNS uses a set of neighborhood structures N_k, with k in [1, k_{max}], and $N_k(S)$ is the set of solutions in the kth Neighborhood of S.

4.7 Shaking Sub-module

In random manner, the shaking function varies and explores new parts of the search space using the neighborhood structure. From an initial solution S, the shaking function chose other solution S' from the kth neighborhood structure $N_k(S)$.

To facilitate this functionality, the VNS algorithm classifies Neighborhood structures in such a way that VNS algorithm increasingly explores further away from the current optimal solution S.

After exploring search space of the current neighborhood structure thoroughly with a local search algorithm, the best local solution, S" is compared with S. such as:

If S" is better than S, it replaces S (s ← S") and the algorithm starts all over again with k = 1. else, k is increased and then the algorithm jumps to the next neighborhood structure (k ← k + 1) and continues shaking phase with this neighborhood structure.

This shaking function stops after visiting all neighborhood structures. The result of each Shake step, S', is used as the starting point for the next Local Search algorithm to generate a locally optimal solution S".

5 Performance Study

5.1 Experimental Environment

To prove the efficiency and good performances of our proposed solution MSND, we implement a testbed environment with a distributed SDN environment based in Mininet emulator using a quad-core machine with Xeon processors 2.00 GHz; 8-core 4 MB; 32 GB of memory. We implement our topologies generator [3], we choose Waxman model [18] as the graph pattern. Our performance studies were performed on a set of 100 random networks topologies. The values of $\alpha = 0.1$ and $\beta = 0.3$ were adopted to generate networks topologies with nodes' average degree is between 3 and 4 in the mathematical model of Waxman.

5.2 Implementation and Results

To demonstrate the performance of our solution MSDN, we compare it with a traditional multicast routing protocol, we use a PIM-SM protocol, and enhanced implementation of PIM-SM protocol with VNS-RP algorithm to select RP router, EDSPA, and BAERA Solutions. Multicast Tree MT cost is measured using Opt_F function defined in formula (4) with $\alpha = 0.3$ and $\beta = 0.7$.

We start our performance studies with Multicast Tree Construction delay metric. Which is the required time to build the all Multicast Tree paths from all sources to all receivers, after receiving all receivers' membership requests sent from all initial receivers by the controller, to construct the multicast tree at initialization of the multicast session, in case of SDN based solutions, and by multicast routing protocol (PIM-SM) in traditional solution. The results of these simulations, as shown in Fig. 5, represent the behavior of the network when the network try to construct the multicast tree. Indeed, we can notice a large difference between SDN based solution and traditional solutions, our solution gives better performance such as it uses, additionally to SDN based solution, a speedy VNS algorithm to construct the multicast tree.

Our multicast session data plan is based on sending a stream of news plan data of type constant bit-rate (CBT) with UDP transport protocol using a command-line tools, and 720p video in a VBR variable bit-rate (VBR) MPEG4 format using (VLC application).

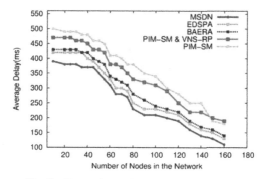

Fig. 5. Comparison of construction tree delay

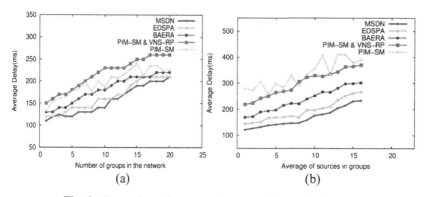

Fig. 6. Comparison of supported groups and supported sources.

Figure 6.a demonstrates an investigation of scalability with respect to supported groups by each proposed algorithm, the networks topologies used for this study comprised of random networks of 100 nodes in around of 10% as multicast group members (3% of sources and 7% of receivers).

Figure 6.b demonstrates an investigation of scalability in respect, this time, to the supported sources in the multicast session by each proposed solution, the networks topologies used for this part of study comprised of random networks of 100 nodes in around of 10% as multicast group members and always just one multicast group.

6 Conclusion

In this paper, we begin our research with an overview of software defined network SDN, multicast IP, and the advantage of integrating SDN in multicast IP technology. We reviewed the multicast SDN proposed in the literature and their algorithmic structures. The multicast tree construction in SDN problem directly affects the structure of the multicast tree and the performance of all the multicast routing scheme accordingly, and consequently the all multicast session. To solve these problems, MSDN is proposed based on as a new multicast controller SDN using VNS heuristic algorithm in the multicast tree construction sub-module. We proposed a mathematical modeling of the cost and delay functions for the multicast tree construction algorithms. Performance results indicate that our proposed solution has good performances in Multicast Tree ST construction delay, all Multicast Tree MT cost, and other aspects. Our future work will be focused on extending this solution to support QoS criteria imposed by multicast receivers across the network, additionally, implementing others multicast modules like group membership in the SDN controller.

References

1. Baddi, Y., Ech-Cherif El Kettani, M.D.: Key management for secure multicast communication: a survey. In: 2013 National Security Days (JNS3). pp. 1–6 IEEE (2013)
2. Karaman, A., Hassanein, H.: Core-selection algorithms in multicast routing—comparative and complexity analysis. Comput. Commun. **29**(8), 998–1014 (2006)
3. Mehlhorn, K.: A faster approximation algorithm for the Steiner problem in graphs. Inf. Process. Lett. **27**(3), 125–128 (1988)
4. Salama, H.F.: Multicast routing for real-time communication of high-speed networks. (1996)
5. Deering, S.E., Cheriton, D.R.: Multicast routing in datagram internetworks and extended LANs. ACM Trans. Comput. Syst. **8**, 85–110 (1990)
6. Hansen, P., et al.: Variable neighborhood search: methods and applications. Ann. OR. **175**(1), 367–407 (2010)
7. Dijkstra, E.W.: A note on two problems in connexion with graphs. Numer. Math. **1**(1), 269–271 (1959)
8. Mckeown, N. et al.: OpenFlow: enabling Innovation in Campus Networks
9. Abbou, A.N. et al.: Software Defined Networks in Internet of Things Integration Security: Challenges and Solutions. In: 2018 6th International Conference on Wireless Networks and Mobile Communications (WINCOM), pp. 1–6 IEEE (2018)
10. Marcondes, C.A.C., et al.: CastFlow: clean-slate multicast approach using in-advance path processing in programmable networks. In: Proceedings—IEEE Symposium Computing Communication, pp. 000094–000101 (2012)
11. Ananta, M.T., et al.: Multicasting with the extended dijkstra' s shortest path algorithm for software defined networking (2014). https://www.semanticscholar.org/paper/Multicasting-with-the-Extended-Dijkstra-'-s-Path-Ananta-Jiang/c75833ae729be310993d77b99fee52be0193df6b
12. Lin, Y.-D., et al.: Scalable multicasting with multiple shared trees in software defined networking. J. Netw. Comput. Appl. **78**, 125–133 (2017)
13. Huang, L.-H., et al.: Scalable and bandwidth-efficient multicast for software-defined networks. In: 2014 IEEE Global Communications Conference, pp. 1890–1896 IEEE (2014)
14. Iyer, A., et al.: Avalanche: data center Multicast using software defined networking. In: IEEE International Conference on Communication System Networks, pp. 1–8 (2014)
15. Baddi, Y., El Kettani, M.D.E.C.: Qos-based parallel GRASP algorithm for RP selection in PIM-SM multicast routing and mobile IPV6. Int. Rev. Comput. Softw. **9**(7), 1271–1281 (2014)
16. Sebbar, A., Zkik, K., Baadi, Y., Boulmalf, M., Ech-Cherif El Kettani, M.D.: Using advanced detection and prevention technique to mitigate threats in SDN architecture. In: 2019 15th IWCMC, 2019, pp. 90–95 (2019)
17. Hansen, P., Mladenovic, N.: Variable neighborhood search: Principles and applications. Eur. J. Oper. Res. **130**(3), 449–467 (2001)
18. Zkik, K., Sebbar, A., Baddi, Y., Boulmalf, M.: Secure multipath mutation SMPM in moving target defense based on SDN (2019)

Author Index